WOMEN IN PHYSICS

WOMEN IN PHYSICS

The IUPAP International Conference
on Women in Physics

Paris, France   7-9 March 2002

EDITORS
Beverly Karplus Hartline
Dongqi Li

Argonne National Laboratory
Argonne, Illinois

SPONSORING ORGANIZATION
International Union of Pure and Applied Physics (IUPAP)

Melville, New York, 2002
AIP CONFERENCE PROCEEDINGS ■ **VOLUME 628**

Editors:

Beverly Karplus Hartline
Deputy Laboratory Director

Dongqi Li
Materials Science Division

Argonne National Laboratory
9700 South Cass Avenue
Argonne, IL 60439
USA

E-mail: bhartline@anl.gov
       dongqi@anl.gov

L.C. Catalog Card No. 2002110176
ISBN 0-7354-0074-1
ISSN 0094-243X

Printed in the United States of America

CONTENTS

PART ONE: RESULTS OF THE CONFERENCE

Discussion Summaries

PART TWO: CONFERENCE PROCEEDINGS

Opening Remarks

International Survey Report

Plenary Papers

Papers by Country

APPENDICES

Preface

The first International Conference on Women in Physics brought together more than 300 physicists—about 15% of them men—from 65 countries for three days of energizing and inspirational presentations, discussion sessions, and informal interactions. Organized by a working group of the International Union of Pure and Applied Physics (IUPAP), the Conference was held at UNESCO headquarters in Paris from 7 to 9 March 2002.

Throughout the world, women are underrepresented in physics and related fields, such as engineering, that require a strong physics background. Moreover, the percentage of women decreases sharply with each step of career advancement. The Working Group on Women in Physics, established in 1999, was charged to understand why, and to develop strategies for increasing women's participation in and impact on the field. It is noteworthy that the Conference brought together physicists from about 50% more countries than are represented in IUPAP.

Under the leadership of Chairperson Marcia Barbosa (Brazil), the Working Group initiated an international survey of women in physics and organized the Conference, inviting teams of physicists from many countries to attend. Each team was asked to develop a brief report and a poster on the situation for women in physics in its country to share at the Conference. Attendees heard the results of the survey and received its preliminary written report. Ten distinguished speakers—at least one from every major geographic region—each provided insights into her own experiences and described the situation, barriers, and actions related to women in physics in her country.

Participants were welcomed by Philippe Busquin, Commissioner for Research of the European Union; Walter Erdelen, Assistant Director-General for Natural Sciences at UNESCO; and Burton Richter, President of IUPAP. Discussions focused on issues and strategies related to six important topics for increasing women's involvement in physics: attracting girls into physics, launching a successful physics career, getting women into physics leadership, improving the institutional climate, learning from regional differences, and balancing family and career.

At its final session, the Conference unanimously adopted eight Resolutions directed at schools, universities, research institutes, industries, scientific societies, national governments, granting agencies, and IUPAP. These Resolutions will be presented at the 2002 IUPAP General Assembly in Berlin. In addition, numerous Recommendations were compiled that feature specific actions or interventions, many of which have been proven successful in one or more countries.

By publishing the Conference Proceedings, the Working Group hopes to expand the impact of the Conference far beyond the participants, each of whom acquired new friends, new colleagues, new perspectives, and renewed commitment to ensuring physics can benefit fully from the ideas and efforts of talented women throughout the world. Readers will find in this volume the welcoming remarks, survey report, invited presentations, ideas and strategies from the discussions, wonderfully diverse country contributions, resolutions, recommendations, sponsor list, and participant list. More details on the Conference can be found at http://www.iupap.org, under a link to the "International Conference on Women in Physics." These Proceedings are available at http://proceedings.aip.org/proceedings/confproceed/628.jsp.

We gratefully thank the 28 organizations that provided funding to make the Conference possible. Counted among the sponsors are several international organizations plus others within individual countries in Asia, Europe, Latin America, and North America—showing the broad geographic range of interest and concern.

Finally, we acknowledge and profusely thank Catherine Kaicher, whose careful editing and formatting prepared the manuscripts for publication; Jacqueline Beamon-Kiene for tireless administrative and logistical support from the time the Working Group was established; Erika Ridgway, who drafted funding requests, handled the finances, and edited much material distributed at the Conference; Sabine Kessler, our contact at the publisher; Kim Quigley for designing the program brochure and logo; Marcelle Rey-Campagnolle, resourceful chairperson of the local organizing committee; and Judy Franz, IUPAP's extraordinary liaison to the Working Group.

Beverly Karplus Hartline and Dongqi Li, Editors
June 2002

PART ONE

Results of the Conference

Conference Resolutions

INTRODUCTION

Physics plays a key role in understanding the world we live in, and physicists contribute strongly to the welfare and economic development of nations. The knowledge and problem-solving skills of physicists are essential in many professions and industries and to society at large. To thrive in today's fast-changing, technological world, every country must achieve a highly educated population of women and men, fully engaged in making decisions important to their well being.

Thus a knowledge of physics is an important part of general literacy for every citizen. In addition, advancing physics understanding is an exciting intellectual challenge that benefits from the diverse and complementary approaches taken by both women and men from many cultures. Currently women can and do contribute to this quest and, through physics, to the welfare of humankind, but only in small numbers: women are an underutilized "intellectual reserve." Only when women participate fully as researchers in the laboratory, as scientific leaders and teachers, and as policy makers will they feel equal partners in a technological society.

The ideas in these resolutions are aimed at bringing more women into the mainstream and leadership of physics. They were unanimously approved by over 300 physicists from 65 countries attending the first International Conference on Women in Physics, held in Paris, France, 7–9 March 2002.

Each country is different. Thus the conference participants are translating these resolutions into their own languages. In the translation, the ideas in the resolutions will be appropriately phrased and directed to the responsible entities in each country.

1. RESOLUTION DIRECTED AT SCHOOLS AND THEIR GOVERNMENT SPONSORS

Girls should be given the same opportunities and encouragement as boys to learn physics in schools. When parents and teachers encourage girls, it strengthens their self-confidence and helps them advance. Methods and textbooks used in teaching physics should include those that have been shown to interest girls in physics and foster their success. Studies show that young girls have a strong desire to help improve people's lives, and therefore it is important that they have the opportunity to see ways that physics has a positive impact on society.

2. RESOLUTIONS DIRECTED AT UNIVERSITIES

2.1 Students

Universities should examine their policies and procedures to ensure that female students are given an opportunity for success that equals that of male students. All policies that perpetuate discrimination should be abolished, and policies that promote inclusion should be adopted. This may involve adopting such practices as: using a broad interdisciplinary approach to physics; providing flexible entry criteria to the physics major; allowing early participation in research; providing mentoring; and exposing students to the important contributions physics makes to other sciences, medicine, industry and the quality of daily life. Adopting these practices will have an especially positive effect on young women, who often feel isolated and unwelcome in physics.

2.2 Faculty and Researchers

Recent studies have shown that, even at top research institutions, women scientists have not been treated fairly with respect to their male colleagues. This is not only very harmful to women in science but in the long run will be harmful to science as well. Universities must examine and communicate their policies and practices to make sure that they promote equity; it is of key importance that universities guarantee transparent and fair mechanisms of

CP628, *Women in Physics: The IUPAP International Conference on Women in Physics,* edited by B. K. Hartline and D. Li
© 2002 American Institute of Physics 0-7354-0074-1/02/$19.00

recruitment and promotion. Additional important elements for success are access to research funding and facilities and sufficient time for research.

Having a family should not be allowed to impede women's participation in scientific careers. A family-friendly environment that provides such things as child-care facilities, flexible working schedules and employment opportunities for dual career families will enable career success.

University governance has been found to be dominated by men. Women need to be included in university and physics department governance, particularly on key policy committees. Women must have input into those policies that control their own destinies. It is important for the development of young women physicists to see successful women active in research, teaching and leadership.

3. RESOLUTION DIRECTED AT RESEARCH INSTITUTES

Research institutes will benefit from policies that allow women scientists to be successful. Institute directors should make sure that policies that promote gender equity in recruitment and promotion are adopted and enforced. Too often what has been termed a "glass ceiling" is allowed to stop the advance of women's careers.

Institute directors should take an active part in ensuring that family-friendly practices such as child-care facilities and flexible working schedules are available to all. Surveys repeatedly show that a leading concern of women is balancing career and family life; having a family should not be allowed to impede successful participation in scientific research.

4. RESOLUTION DIRECTED AT INDUSTRIES

Industries will benefit from policies that allow women scientists to be successful. Industrial managers and research directors should make sure that policies that promote gender equity in recruitment and promotion are adopted and enforced. Too often what has been termed a "glass ceiling" is allowed to stop the advance of women's careers.

Industrial managers should take an active part in ensuring that family-friendly practices such as child-care facilities and flexible working schedules are available to all. Surveys repeatedly show that a leading concern for women is balancing career and family life; having a family should not be allowed to impede successful participation in scientific research.

5. RESOLUTION DIRECTED AT SCIENTIFIC SOCIETIES

Scientific and professional societies can and should play a major role in increasing the number and success of women in physics. Each society should have a committee or working group that is responsible for such issues and that makes recommendations to the society as a whole. At a minimum societies should do the following things: work with other organizations to collect and make available statistical data on the participation of women in physics at all levels; identify women physicists and publicize them as role models; include women on program committees and as invited speakers for society-sponsored meetings and conferences; and include women on editorial boards of society journals.

6. RESOLUTION DIRECTED AT NATIONAL GOVERNMENTS

Physics plays a key role in understanding the world we live in, and physicists contribute strongly to the economic and cultural development and welfare of nations. It is therefore in every nation's self-interest to provide strong physics education for all its citizens and to support advanced education and research. Governments must ensure that women have the same access and chance for success in research and education as men. National planning and review committees should include women, and awards of government funds should only be made to organizations and institutions that make gender equity a part of their policies.

7. RESOLUTION DIRECTED AT GRANTING AGENCIES

Agencies that make funding available for scientific research play a key role in promoting the success of individual scientists as well as science as a whole. Past studies have shown evidence for gender bias in the review process. Therefore, to ensure that women have the same access to research funding as men, all competitions for funding should be transparent and widely publicized; the criteria for obtaining funds should be clear; and women should be included on all review and decision making committees. Limits on age of eligibility or grant structure and duration that seriously disadvantage applicants taking family leave should be reconsidered. Granting agencies should maintain and make available statistical data by gender, including such information as the proportion and qualifications of women and men who apply for funding and who obtain funding.

8. RESOLUTION DIRECTED AT IUPAP

IUPAP is the international organization of physicists and as such exerts considerable influence on the physics community through its statements and activities. IUPAP should both endorse the above resolutions aimed at other groups and also examine its own actions to make sure that they contribute to increasing the number and success of women in physics. It will also be valuable for IUPAP to communicate the results of this conference to international scientific organizations in other fields. In the election of IUPAP's Executive Council and Commission members, procedures should be instituted to ensure the full inclusion of women. IUPAP sponsors major international conferences; a criterion for such sponsorship should be the demonstration that women are included on the International Advisory Committees and Program Committees. IUPAP should require conference organizers to report gender distribution of invited speakers. IUPAP should encourage all of its national Liaison committees to include women among their members. Liaison committees should also advocate these resolutions in their countries. IUPAP should continue its Working Group on Women in Physics and empower it to establish an international advisory committee with a member in as many countries as possible. Finally, this group will form the basis of a network that can continue the work of increasing the number and success of women in physics.

Discussion Summaries

Top: Quite a few infants and children
attended; in most cases they visited Paris in
the care of their fathers

Above: Local organizers
arranged a great dinner
cruise on the Seine Thursday
evening, where physics
networking continued and
more international
friendships formed.

In small group discussions each afternoon, participants shared experiences, insights, and solutions aimed
at six topics key to increasing the presence and impact of women in physics (see Topic Summaries).

Topic 1: Attracting Girls Into Physics

Sumathi Rao[1*], Jenni Adams[2], Aba Andam[3], Åshild Fredriksen[4], Neelima Gupte[5], Jyoti Gyanchandani[6], Christa Hooijer[7], John O'Brien[8], and Peter Saeta[9]

[1]Harish-Chandra Research Institute, India; [2]University of Canterbury, New Zealand; [3]University of Ghana; [4]University of Tromsø, Norway; [5]Indian Institute of Technology, Madras; [6]Bhabha Atomic Research Centre, India; [7]Foundation for Fundamental Research on Matter (FOM), the Netherlands; [8]University of Limerick, Ireland; [9]Harvey Mudd College, USA

Abstract. We discuss the problems in attracting girls into physics. We present some of the projects undertaken in various countries to ameliorate the problem, and we conclude with some follow-up suggestions that can be implemented in all countries.

THE PROBLEMS

There appeared to be three main problems in attracting girls into physics, which seemed to cut across countries, even across the developing/developed country divide. These problems are the image of physics, the quality of physics teaching in schools, and societal expectations of girls and women.

Image of Physics and Physicists

The first problem is the image of physics and physicists. As a subject, physics is perceived as difficult and boring, and as people, physicists are perceived as mostly male, "nerdy," and dull, at least in most developed countries. In the Third World, on the other hand, there is no perception at all of what physicists, or even scientists, do. Society only recognizes professions like medicine, engineering, and law. Hence, employment prospects for physicists (other than in academics or teaching) are practically nonexistent.

Physics Teaching

Common to all countries is the fact that kids have no real idea of the value of doing physics, and their entire experience of physics is through their physics teacher. This leads us naturally to the second problem, which is that of physics teaching in schools. Here, the main problem is the lack of enthusiastic and motivated teachers who themselves understand physics.

But for girls in particular, it was felt that the classroom atmosphere and examples used are particularly alienating. Specific problems mentioned were that many girls' schools (particularly in the Third World) have no labs or hands-on teaching, with the result that girls, later in college, are "frightened" of experiments. Textbooks are often of poor quality and illustrations show more boys or unsexed figures than girls, which is especially unfortunate because children need to identify with the illustrations.

Ways of teaching do not take into account the fact that many girls learn better in collaborative groups than in a competitive atmosphere. Finally, questions were also raised about whether introduction to abstract concepts is poorly timed for girls, i.e., that girls languish in dull courses early on waiting for boys to "grow up."

---

* Topic chairperson and corresponding author: sumathirao1999@yahoo.com

CP628, *Women in Physics: The IUPAP International Conference on Women in Physics,* edited by B. K. Hartline and D. Li

Societal Expectations

The third problem is societal expectations. These are far more acute in Third World countries, where girls' education is not taken seriously. The general philosophy is, "why pay more to have girls study physics or even science in general, when they are going to drop out and get married anyway?" In many countries, this leads to girls themselves not taking their work very seriously because they always have the fall-back option of getting married. But even in developed countries like Israel, a similar philosophy means that programs for gifted children are encouraged less for girls than for boys.

ATTEMPTS TO SOLVE THE PROBLEMS

Now, how are the various countries trying to solve these problems? In UK, Ireland and Scotland, where they face an acute shortage of physicists, several programs have been initiated where professors and postdoctoral fellows from universities give lectures to school children, even at the elementary-school level. They have found that there is an enormous impact on young girls when the physicist speaking is a woman. They have also initiated programs in which school students are taken to visit laboratories and universities and interact with physicists. The universities have also encouraged this interaction by recognizing this kind of work as part of the work of the scientist by giving it due consideration in promotions and awards. The scientist can even use a small part of her/his grant money for this purpose. This effort seems to have had a positive effect on enrollment of students in physics in some parts of the country.

Other countries like Ghana have had science clinics where they have worked on changing boys' attitudes toward girls in high schools. Brazil has experimented with having special classes to improve scientific skills oriented mainly to women students. In South Africa, physics graduate students offer to solve problems for industry in an effort to make the industries realize the value of physics. This is perhaps an experiment that should be tried in other Third World countries.

RECOMMENDATIONS

Physics Teaching

What more can and should be done in all countries?
1. Without question, the first thing is to improve the quality of physics teaching. This would involve various steps. One needs to provide money for science teaching and well-equipped laboratories.
2. Teaching should be valued, which in turn means that teachers should be better paid.
3. Teachers need to be trained and should be knowledgeable about current physics. Therefore, money is needed for teacher training programs, and for interaction with working physicists.
4. Teachers should be educated in gender issues, and the classroom atmosphere should be "girl friendly." For instance, girls' toys can be used to illustrate physics principles. Textbooks for young children should include figures of both men and women.
5. The curriculum should be revised to make it more interesting to girls by relating physics to its applications to technology, to making the world a better place, to the environment, and so on. The textbooks used should consciously include contributions of women physicists, who can then serve as role models.

Image of Physics and Physicists

The media should be used effectively to popularize physics, by showing what physicists (men and women) really do. There is a need to change the "nerdy" image of physics by showing different types of physicists doing different jobs. This will teach society that the skills one learns as a physicist are transferable and that there are many job prospects for physicists. Perhaps this will also influence industries in Third World countries to look at physicists as useful problem solvers.

Role Modeling

It is very important everywhere to improve interaction between schools and universities (or laboratories) where physics is actually practiced. This will give young students hands-on experience in research. It will also provide role models, particularly for young girls to see women physicists. (But this interaction needs to be monitored, because not all scientists are good communicators or good motivators.)

Universities need to support and reward men and women who engage in efforts to get boys and girls into physics, but in particular those who make special efforts to attract girls. Women physicists have an important and special role to play. They should be encouraged to work with young girls in schools, because personal contact is very important. The total numbers are so low that even if each woman physicist attracts one more woman into physics, it will be significant!

There is a general perception that competition puts off girls; however, it would be a good idea to encourage more girls to enter international competitions in physics. For this to be successful, a "critical mass" of girls is needed, because girls need peers. One suggestion was to network the smart girls and form some kind of "smart girls club." Other suggestions were to have special incentive scholarships for girls, have special awards and recognitions for women in physics, and provide mentoring programs for young girls in physics.

Other Suggestions

It may also be necessary to shorten the post-master's-degree route (Ph.D., postdoctoral fellowships) to a permanent academic position in physics, to attract girls (and even boys) into physics in larger numbers.

Some of the other suggestions involved specific problems in different countries. Countries with too early specialization into science and humanities need to have more flexible undergraduate programs. To overcome societal biases in several countries, counseling to parents, teachers, and career advisors is needed to encourage girls into physics. In other words, the social climate itself in some countries needs to be changed—a daunting task!

But the conclusion was that all countries felt that much more needs to be done to attract more girls into physics, and that the resolutions at this conference will help in that effort.

Topic 2: Launching A Successful Physics Career

Beverly Karplus Hartline[1][*], Engin Arik[2], Marilia Caldas[3],
Gillian Gehring[4], Dimitra Darambara[5], Annalisa Fasolino[6],
Liv Hornekaer[7], and Peter Melville[8]

[1]*Argonne National Laboratory, USA;* [2]*Bogazici University, Turkey;* [3]*University of Sao Paulo, Brazil;*
[4]*Sheffield University, UK;* [5]*University College London, UK;* [6]*University of Nijmegen, Netherlands;*
[7]*University of Southern Denmark;* [8]*Institute of Physics, UK*

Abstract. The facts that there are few women in physics and their presence decreases at each higher level in the hierarchy are evidence that women physicists face challenges in launching and pursuing a successful physics career. In four parallel discussion groups led and recorded by the coauthors, about 80 of the conference participants discussed the issues involved, shared proven solutions, and brainstormed other possible strategies. This paper summarizes the results of the discussions, emphasizing useful advice for aspiring women physicists and those interested in facilitating their retention and advancement.

ISSUES

Background

Physics knowledge and skills are assets for careers in academia, industry, government, finance, writing, and even medicine. In these various arenas, physicists perform research, teaching, management, or service. Some careers require education to the doctorate level, while others require only bachelor's or master's degrees. In many countries, one must have a doctorate in physics to be considered a physicist; thus, most conference attendees either had physics doctorates or were currently enrolled in a doctoral program. Whereas women comprise a minority of physics Ph.D. recipients in all countries, ranging from a few percent to nearly 30%, 10 years after the degree, the proportion of women in the physics workforce in every country is substantially lower. Each attendee knows talented colleagues— both men and women—who gave up physics. But the fraction of women who have done so is much larger than the fraction of men.

Defining Success

To the discussion participants, success centered on being able to make their best contributions to physics in a subfield and career path of their interest and choosing. Beyond this essential element, success had three major dimensions: earning *recognition* within the physics community, making an *impact* on society, and achieving *personal goals. Recognition* involves advancement within the hierarchy, the respect of peers and students, speaking invitations, job offers, awards, funding, and salary. *Impact* relates to improving the world, one's country, or the profession of physics. *Personal* success factors could involve independence, fun, life balance, or rewards commensurate with effort. Individuals place different importance on the three dimensions; moreover, typically the relative prominence of each dimension changes during a person's career.

---

[*] Topic chairperson and corresponding author: bhartline@anl.gov

CP628, *Women in Physics: The IUPAP International Conference on Women in Physics,* edited by B. K. Hartline and D. Li
© 2002 American Institute of Physics 0-7354-0074-1/02/$19.00

Career Phases and Issues

During and immediately after graduate school, the focus is on choosing a physics specialty and career path (e.g., research, teaching, industry, government). It is important for students and recent doctorates to think about and set near-term and long-range goals, and simultaneously to be flexible and alert to exciting but unanticipated opportunities that might arise. Key challenges for those pursuing a research career, for example, are to obtain funding and a position and to earn high marks from peer reviewers assessing research proposals. Often, during the early-career phase the demands of family responsibilities (marriage, infants, and young children) are particularly strong and divert women physicists from the single-minded career focus expected by employers. Many of the successful senior women physicists at the conference managed to raise children. It is important that women, men, and employers realize that women can have both a physics career and a family (in most countries).

To succeed professionally, one's reputation and visibility must grow. Both in graduate school and in the workplace, good supervisors or mentors can be very helpful. Women whose professors and first employers actively championed their physics potential and performance gained valuable self-confidence. Those who were given important assignments advanced comparatively rapidly. Doing a great job on a minor or invisible assignment is clearly better than doing a miserable job in it—but it does not help one's career as much as success in an important and visible assignment. Opportunities to lead even a small subproject typically provide a better career boost than always being a follower. Discussants suggested that women participating in enormous collaborations (such as those common in experimental high-energy physics) need to ensure they are not limited to support roles and contributions.

A woman who seeks promotions, professional recognition, and rewards must understand the "rules of the game," and deliberately acquire the skills and experience needed to enter the next higher level. These "rules" are the collection of practices, behaviors, and expectations—often unwritten—that determine who gets hired or promoted. In most countries, the rules of the game in physics are not particularly clear, and they produce barriers that few women physicists have surmounted. In some cases, for example, the "rules" seem to favor highly competitive behavior, called "combat physics" by conference participants, that discourages and even offends many women.

Participants have mixed feelings about the value of affirmative action regulations. In some instances these requirements are necessary to overcome clear barriers to women. In other cases they involve inflexible procedures and carry a reverse-discrimination stigma, which perversely makes it even harder for women to advance.

IMPORTANT INTERNATIONAL DIFFERENCES

The 65 countries represented at the conference have a broad range of social/cultural expectations of women and women's roles in society. In some cultures the education of women is not a high priority. In others, family responsibilities make it nearly impossible for women to pursue any career, much less one as demanding as physics.

The size of a country and its physics enterprise proves to be an important difference during the early career. In countries with a small physics community, it is essential for aspiring physicists to spend several years abroad. This experience helps the physicist develop the professional contacts, collaborators, credentials, and professional reputation needed to be selected for a permanent position in the home country.

The affluence of a country determines whether it has funding opportunities or an infrastructure for physics research. In poorer countries, experimental specialties are simply not viable because the necessary apparatus is unavailable and unaffordable.

Perhaps the most important differences are between developed and developing countries. Developing countries usually have less advanced schools and have few, if any, universities. If there is a university, it is not among the best in the world, thereby disadvantaging those educated there. Developing countries typically have no research opportunities. If there is significant industry, it usually exploits low-wage manual labor, rather than offering high-tech, intellectually challenging jobs.

SUCCESSFUL SOLUTIONS

Successful women physicists in every country are extraordinarily talented and very passionate about physics. These women have had to take a lot of initiative, to seek out people—especially leading physicists—for advice, to ask for what they need or want, to promise a lot and accomplish more, and to overcome adversity of many types. Self-confidence is essential: when families, teachers, advisors, and colleagues give girls and women strong encouragement and praise their accomplishments and contributions to physics, self-confidence rises. Many successful women research physicists have chosen to be pioneers in a comparatively new subfield, where perspectives and approaches have not yet become as rigid as they are in established specialties. At institutions where men and women in leadership positions have noticed and remedied inequities and "leveled the playing field," women physicists are growing in number and prominence.

RECOMMENDATIONS

The following recommendations emerged in the discussions about launching a successful career. Several are based on strategies that have proven successful in one or more countries. Others are promising ideas developed by conference participants. Most have been incorporated in the Conference Resolutions and Recommendations. Not all will be applicable or useful in every country or situation.

- Create, support, and encourage networks for women physicists—local, national, and international, including a worldwide e-network—to provide career support, advice, role models, and information about job opportunities and funding sources. Networks are proven means to help women physicists help each other and overcome the marginalization that often occurs when they are the lone woman in a group.
- Have transparent, gender-blind processes for important decision making. To improve accountability, require that important decisions be reported and explained. Important decisions include those related to recruiting, promotion, salaries, funding, and the allocation of space and equipment.
- Place women on important committees and policy-making bodies to improve the system, change the "rules," and give the women opportunities to have visible impact.
- Provide early and frequent opportunities for women to participate in physics research, author or co-author papers, attend conferences, and give physics talks.
- Provide enlightened and supportive supervisors and mentors for women physicists at every career stage. These people should treat the women in a respectful and collegial manner, obtain funding, teach the women the "rules of the game" and how to write successful proposals, introduce them to important professional contacts, give them challenging assignments and opportunities, provide constructive feedback on unsuccessful proposals or interviews, give them credit, send them to conferences, and advocate them in the physics community.
- Provide training for women physicists in presentation of results, paper writing, job applications, grant writing, the "rules of the game," assertiveness, etc.
- Shorten the post-post-doc phase with its insecurity, relocation requirements, and tendency to create a "two-body" problem for couples that form during this period.
- Establish a speaker program for women physicists. Maintain a web list of speakers and topics, and provide a funding source(s) to support the costs of speaker travel.
- Organize prestigious, topical international physics summer schools with significant participation and leadership from women physicists and convenient child care.
- Ensure that national and international sources of funding for scientific research give women an equitable opportunity for awards.
- Accord a career interruption for "family service" the same respect as military service gets in most countries. Pause the career clock, have flexible age limits for grants and fellowships, and provide resources to help women restart their careers after a break.
- Help dual-career couples find two appropriate positions in close proximity.
- Collect and publish data internationally on physics demographics, including gender, to watch and influence trends in the participation and advancement of women. Also collect data on why women leave physics.

SUMMARY

Women have insights, ideas, and much energy to contribute to physics. Yet in essentially all countries many talented women are lost from the physics profession between graduate school and a position of leadership. If implemented, the ideas and recommendations developed at the conference and particularly in the discussions on "Launching a Successful Physics Career" should help increase the retention, visibility, and advancement of women in physics.

Topic 3: Getting Women Into Positions of Leadership Nationally and Internationally

Katharine B. Gebbie[1*], Azam Iraji-zad[2], Hélène M. van Pinxteren[3],
Kimberly S. Budil[4], Jo Ann C. Joselyn[5], and Laurie E. McNeil[6]

[1]National Institute of Standards and Technology, USA; [2]Sharif University of Technology, Iran;
[3]Foundation for Fundamental Research on Matter (FOM), The Netherlands; [4]Lawrence Livermore National
Laboratory, USA; [5]International Union of Geodesy and Geophysics, USA; [6]University of North Carolina, USA

Abstract. The paucity of women physicists in positions of influence worldwide has implications for industry, government, and academia, as well as for the future of the profession itself. Women from more than 45 countries discussed the importance of having women in leadership positions and shared their experiences and successes. While implementation will differ among countries, we report a set of recommendations addressing the preparation of women for leadership, the selection process, and the responsibilities of institutions.

THE ISSUE

While the nature and magnitude of the underrepresentation of women in physics varies from country to country, the percentage of women in physics in all countries decreases markedly with each step up the academic ladder and with each level of promotion in industrial and national laboratories. Nor are women well represented among physicists in top research institutes, funding agencies, professional societies, or government. The result is a dearth of women among physicists in positions of leadership worldwide.

Why does this matter? What can we do to increase the numbers? In attempting to answer these questions, we identify positions of leadership not simply as the opportunities for exercising leadership that occur at all stages of a career, but rather as formal positions with control over allocation of human and financial resources and the research agenda.

The participants in the discussions felt that women should have such positions not so much because it is fair, which it is, but because women represent a largely untapped source of talent and innovation. If the profession is to serve its goal of advancing and diffusing the knowledge of physics, it must draw upon the widest possible spectrum of talented individuals—the best and the brightest of both sexes and all segments of society. For the profession to thrive, women must feel that they have opportunities for advancement and influence.

And just as multidisciplinary teams can be exceptionally innovative, so bi-gender teams can bring a rich diversity of approaches to decision making. Because decisions regarding resource allocation and programmatic direction affect female and male physicists alike, females as well as males must be involved in making them. No longer can the profession of physics afford to limit the perspective and judgment of its leadership to those of a single gender.

If science and technology are to fulfill their potential to improve the quality of life in all countries, and indeed if democracy is to thrive, we must make it our goal to achieve a scientifically literate society, a population that understands and values the contributions that science can make to our well being. Women are half that population. Only when women see that women are participating fully in the scientific endeavor—as scientific leaders and policy makers as well as researchers in the laboratory—will they feel equal partners in a technological society.

---

* Topic chairperson and corresponding author: kgebbie@nist.gov

CP628, *Women in Physics: The IUPAP International Conference on Women in Physics,* edited by B. K. Hartline and D. Li
© 2002 American Institute of Physics 0-7354-0074-1/02/$19.00

NATIONAL DIFFERENCES

Because cultural, social, and political factors all have roles in the careers of women physicists, each society must develop its own guidelines for enhancing the status of women in physics. Some of the differences among societies are discussed in the report of Topic 2 in these Proceedings, *Launching a Successful Physics Career*. The issues around getting women into positions of leadership are in some ways different from those involved in launching a career and are less easily associated with the size, affluence, or developmental stage of a country.

In some countries, the pipeline is blocked by lack of opportunities for advancement; in others the pipeline is empty. In some countries, women who reach top positions command as much or more respect than their male peers; in others men do not feel comfortable dealing with women or taking orders from them. In some countries it is essential for career advancement to have studied abroad; in others it is detrimental to have done so. In several countries the numbers of women on high-level committees is increasing, but the women are often selected for their willingness to follow rather than their ability to lead.

In the remainder of this report, we outline recommendations that emerged from our three discussion groups composed of women physicists from more than 45 countries. Many of our recommendations appear in the reports of other discussion groups; most have been incorporated into the Conference Resolutions and Recommendations.

RECOMMENDATIONS

We group our recommendations for getting women into positions of leadership under three headings: preparing for leadership, the selection process, and responsibilities of institutions.

Preparing for Leadership

- Men in positions of leadership in physics must be held accountable for making change happen—for creating a climate in which every physicist can contribute to the full extent of his or her ability. This responsibility extends not only to the preparation of women for positions of leadership but also to the selection process, and above all, to the results.

- Women in positions of leadership in physics have a special responsibility to demonstrate their commitment by mentoring, supporting, promoting, and creating opportunities for other women; drawing attention to policies, processes, and attitudes that are discriminatory toward women and working to change them; and acting as role models and mentors for students, including very young women, by actions such as visiting schools, presenting talks and demonstrations, and inviting students to visit their laboratories.

- Younger women physicists for their part must take responsibility for preparing themselves for leadership by seeking out mentors, seeking out opportunities to gain leadership experience and learning how to qualify for them, learning about the culture and informal "rules" of their institutions, and supporting each other.

The Selection Process

- The criteria and decision-making processes for appointments, professional advancement, awards, etc., must be made public and transparent. In job searches, the final choice, the ranking, and the justification thereof must be available to all candidates.

- Review boards for hiring, promotion, and funding must include women.

- Equity must be judged not by qualitative assessment of the process, but by quantitative measure of the outcome. If the outcome does not reflect the pool of available candidates, the process must be inferred to be biased.

Responsibilities of Institutions

Industry, Government, Academia

- Establish flexible management structures, e.g., shared leadership positions that take advantage of a diversity of views and allow scientist-managers to gain a broader perspective of their institutions while retaining their research base.
- Establish flexible career models that allow for careers that ebb and flow with different amounts of time devoted to work, family, and community at different times of life. Specifically, such models could allow for the possibility of gaps in careers, of part-time work schedules, and of working at home.
- Provide formal training for leadership positions, e.g., training in speaking, writing papers, and preparing grant proposals. While these skills are equally important for men and women, opportunities for women to acquire these skills may be harder to come by.
- Provide funding for women to attend conferences, particularly international conferences, and nominate women to give invited talks.
- Ensure that women are represented on all hiring and promotion committees.

Professional Societies (e.g., physical societies and national academies)

- Create national and international networks based on this conference. While professional societies can most effectively formalize, monitor, and support such networks, each of us can and should participate.
- Create lobbying groups for women's issues.
- Ensure that women, including students and postdocs, have the opportunity to present talks, chair sessions, and serve on committees, all of which provide both the training and visibility necessary to assume positions of leadership.
- Create a roster of women physicists, nationally and regionally, who are available to present talks and to serve on committees.

Funding Agencies and Evaluation Boards

- Include women on peer review committees and take into account the climate for women scientists when making appraisals.

FOLLOWUP ACTIONS FOR CONFERENCE ATTENDEES

- Establish national and international support systems, e.g., direct mentoring, e-mail support, networking, and workshops to help women physicists share experiences and research facilities. This is important for all women physicists, but particularly for those in countries where women are at particular disadvantage.
- Collect data on the ratio of men and women in top management positions in leading universities, research institutes, professional societies, and funding agencies. Data are important not to focus on past discrepancies or precise statistics, but rather to demonstrate that there is a problem and to establish a baseline against which to measure future progress.
- Actively pursue invitations to speak about this conference at schools, universities, and conferences.
- Urge professional societies to establish committees, websites, and electronic bulletin boards for women.

CONCLUSION

Change will not happen quickly, and it will not happen by itself. Vigorous efforts to increase the numbers of women entering physics can produce rapid—and sometimes short-term—results. Getting these same women into positions of leadership will require a determined, unrelenting effort on the part of men, women, and institutions. In keeping with an Iranian adage, we must move wisely, smoothly, effectively, like a swimming duck.

Topic 4: Improving the Institutional Structure and Climate for Women in Physics

Ling-An Wu[1]*, Manjula Sharma[2], Larissa Svirina[3], Joanne Baker[4], Anne Borg[5], and Peggy Fredrickx[6]

[1]*Institute of Physics, Chinese Academy of Sciences;* [2]*University of Sydney, Australia;* [3]*Institute of Physics, National Academy of Sciences of Belarus;* [4]*Oxford University, UK;* [5]*Norwegian University of Science and Technology, Trondheim;* [6]*University of Antwerp, Belgium*

Abstract. It is well known that in physics the ratio of women to men decreases dramatically up the academic ladder. For the relatively few women who succeed in finding a physics career, the institutional environment is often not conducive to professional advancement, and may even be hostile. About 60 conference delegates in three group sessions discussed the problems encountered in various countries. We summarize the issues raised and recommendations for improving the institutional structure and climate for women in physics.

THE ISSUES

The issues raised covered two broad areas: (1) "hardware" issues such as office and lab space, experimental equipment, child-care centers, and restrooms, and (2) the less tangible issues of safety, discrimination, and sexual harassment that can make the work environment unpleasant for women.

First, the following questions were asked of everyone at the session: Do women at university and research centers have the same amount of office and lab space as their male colleagues? Are child-care facilities available at, or close to, the workplace? Are there the same numbers of restroom facilities for women and men? Is it safe to work late at night and on weekends? Are there disguised institutional discriminatory policies, e.g., regarding age? Are there any policies regarding sexual harassment?

Of course, in the course of discussion other points were uncovered, some of which were specific to a particular country, and some that were also relevant to the other five conference discussion topics. Below we summarize the main issues raised and our recommendations for improving the institutional environments of women in physics.

Work Space and Resources

In general, women have less office and lab space than their male counterparts. This is not merely because of their lower positions but because they are isolated from the "Old Boys' Club," and therefore left out of the decision making that often occurs during informal meetings of the male-dominant leadership. Women are also not as vociferous as men about claiming their due rights. For example, a woman physicist in Denmark had to share her husband's office. There is also the incredible case of the Massachusetts Institute of Technology (MIT) study in the United States, which revealed that the supposed work area given to women professors actually included the area of adjacent corridors, thus inflating the statistics by 20%. In developing countries, where physics itself is underrepresented and jobs are few, it is even more difficult for women to gain their fair share of resource allocation.

---

* Topic chairperson and corresponding author: wula@aphy.iphy.ac.cn

CP628, *Women in Physics: The IUPAP International Conference on Women in Physics,* edited by B. K. Hartline and D. Li

Living Conditions

Living conditions are often inadequate for women in physics institutions in certain countries—possibly for historical reasons because women have always been a minority—and female visitors are treated with indifference. In Antarctica, South Africa, and Brookhaven Laboratory in the United States, there are mixed dormitories or apartments, making life uncomfortable for both sexes. In Denmark and Germany there are mixed toilets, while at MIT there is altogether a serious lack of toilets for women. In one institution in Denmark there are showers for men but not for women.

Child-care facilities for children are nonexistent or insufficient in many developed and developing countries, including Japan, Germany, Belarus, Malaysia and Egypt. In some universities the day care center only accepts the children of professors, and not of students. Mainland China has good child-care facilities, including weeklong boarding nurseries, but they are now decreasing in number due to the decrease in birth rate.

Safety and Harassment

Safety in labs and while working at night or on weekends is generally good, though instruction with regard to radiation and other workplace hazards could be improved.

The main problem raised was sexual harassment. This seems to be more culture- than economy-related, because it is prevalent in Japan, South Africa, and the Western and Latin American countries, despite specific laws. Extreme cases, though, are rare. It exists in various forms, such as pornographic posters and screensavers, sexual slanders, derogatory allusions to young women lecturers, patronizing attitudes, and affairs with female students by male professors (common in Latin America). Such open harassment is unheard of in China and Islamic countries like Iran and Egypt.

Gender Discrimination

Gender discrimination occurs in all countries, even though it may be in a subtle, imperceptible guise. Most countries have laws defining equity for men and women, but implementation is far from ideal. In the UK, women professors still get paid less for the same amount of work. Allocation of resources is also a serious problem, with hidden criteria only known to those in the power ranks, so that the difference is as much as 60% in some countries. In Egypt mobility is restricted for women, who thus cannot lobby directly with male officials.

A major issue is age discrimination, which sounds non-gender-related but in fact is strongest against women. Because a woman has to spend several vital years raising a family, the peak period of her academic productivity is necessarily postponed, and this will weigh against her when she applies for research grants and promotions. She will, of course, be ineligible for awards restricted to investigators below a certain age, and in applying for a job cannot compete with a man who has the same number of publications but is younger. The 12-year limit for pretenure positions in Germany thus discriminates against women, as do recently introduced preferential policies for young scientists in mainland China, where the retirement age is also earlier for women, although women live longer. In Denmark, moreover, there is an unofficial policy that women over 45 are not given research grants. However, affirmative action has been taken in the U.S., UK, the Netherlands, and Canada, where special grants or positions have been established for women.

Even where anti-discrimination laws are in force, there is often insufficient recognition given to women, with the credit all being given to the men. The secretary image of women is ubiquitous, and many a woman physicist has been mistaken for the secretary (when met in person) or assumed to be a man (when known only from her scientific papers). Male dominance is accepted: in most countries a wife adopts her husband's surname, which leads to the problem of what name to use on publications. A typical case occurred in Belgium, where a university refused to allow a woman physicist to use her maiden name in papers she authored. In Japan, male professors resent having more female faculty and do not want to be advisors to women students.

Old, large institutions tend to be more conservative, and any challenge to authority causes great reaction. Women in leadership are viewed with antagonism and face long-standing prejudices: if they are strong they are perceived as too assertive; if they are modest they are seen as too weak. Those who do make it to the top are often burdened with more than their fair share of the teaching or administrative load, to the detriment of their research. A successful solution was found in Belgium, where a department set up a three-person leadership; each co-leader benefited from a much-reduced administrative workload.

In general, there were some marked regional differences for the problems identified. For developing countries, where work conditions and resources lag far behind those of developed countries, it is even more difficult for women to obtain their fair share of the available facilities. In developed countries, sexual harassment seems to be more acute. However, the subtle dominance of the "Old Boys' Network" is a problem common to all countries.

RECOMMENDATIONS FOR IMPROVEMENT

1. Anti-discrimination laws should be enforced and affirmative actions adopted where possible, though the latter would depend on regional differences. State legislation is not enough, there must be measures to ensure implementation. Each institution should recognize that there are problems and should formulate "equal policy rules" forbidding all forms of discrimination related to gender, age, children, etc. Guidelines defining what constitutes discrimination and harassment should be drawn up in accordance with regional distinctions. Regional physical societies should also participate in the formulation of such policies. These rules should be made clearly known to all.

2. Transparency should be improved at all decision-making levels and the traditional dominance of the "old boys" eliminated, especially with regard to hiring, salaries, promotions, grants, and resource allocation.

3. Independent accredited monitoring committees should be set up (like the five-person committee in the U.S.), which would be approved and possibly funded by the regional physical societies. The committees would investigate the allocation of resources and grants, salary levels, and working conditions for women at the institutional level, and review the implementation of anti-discriminatory policies.

4. Networking and interaction among women should be enhanced, both horizontally (with peers) and vertically (among senior and junior faculty and students). Women should actively support one another, especially their younger colleagues. An institutional council (which should include men) should be set up where women can discuss problems and obtain counseling. Women should have the courage to speak out and to make themselves more visible.

5. The work environment should be more family friendly, with day-care centers for children (and possibly elderly dependents) available to all, decent restrooms and dormitories, and appropriate nonsmoking enforcement.

6. Training courses should be organized for both male and female staff. For women, advice should be given on how to have self-confidence, avoid pitfalls, learn leadership skills, and make men allies. Men should be encouraged to treat women as colleagues, trained to recognize what conduct is discriminatory and may constitute harassment, and made aware that women face special difficulties (as, indeed, their own daughters will in the future).

7. The overall situation should be reviewed every year at national physical society meetings with full female representation, and appropriate actions adopted.

Topic 5: Learning From Regional Differences

Pia Thörngren Engblom[1*], Karen Janssens[2], Ann Marks[3], Sue McGrath[4], Elsa Molinari[5], Sharon Stephenson[6], Ulla Tengblad[1], and KarolineWiesner[1]

[1]Uppsala University, Sweden; [2]University of Antwerp, Belgium; [3]Women in Physics Group, Institute of Physics, UK; [4]W5 Discovery Centre, Belfast, Ireland ; [5]INFM and University of Modena and Reggio Emilia, Italy; [6]Gettysburg College, USA

Abstract. There is a widespread notion that there are more women physicists in some parts of the world, such as in southern Europe. We address the questions of to what extent this is true and for what reasons women are more successful in climbing the academic ladder in certain countries. We summarize the discussion sessions, present the recommendations that emerged, and suggest what can be learned from regional differences.

To assess what the obstacles are for the advancement of women in science in general and in the physics community in particular, it is valuable to know what the regional differences and similarities are. The goal is not to present a list of the best countries, but to find effective means to embark on a road for gender equity everywhere. A global perspective is necessary today, and learning what works to improve conditions in one country will serve as an inspiration for women physicists all over the world.

About 80 conference attendees from 42 countries participated in the discussion. We divided ourselves into four subgroups led and recorded by the coauthors. Because, to our knowledge, little has been documented comparing conditions for women physicists in different countries, we surveyed the team leaders of a few representative countries by email before the conference. The intention was merely to gather some background material for the discussion leaders to transmit to the discussion subgroups formed at the conference. Other sources of information were the ETAN report (http://www.cordis.lu/improving/women/documents.htm), based on data from the European Union member countries; data about development over time from the American Institute of Physics; and, of course, the conference attendees themselves. The importance of gathering consistent demographic data on women physicists over time was stressed in several of the topic discussion summaries. Conditions in different countries, such as the deterioration of the socioeconomic status of women in the newly independent states, must be taken into account.

DISCUSSION THEMES

What Are the Regional Differences?

Figure 1 shows the current participation of women physicists at different levels of academia, as estimated by team leaders. The accuracy of the numbers is thus uncertain; still, some general trends can be extracted. In Europe we found an asymmetry favoring the southeast of Europe, while there are fewer women physicists in the northern and western parts of the region.

This trend extends into Turkey, where 35.3% of bachelor's degrees in physics were received by females from 1982 to 1990. For the same period, 36.2% of all bachelor's degrees in Turkey were awarded to women, whereas for the same time span in the United States, 51% women earned a bachelor's degree but only 13% of those who gained physics degrees were women. By 1998 this number had increased somewhat to 19% in the U.S.

---

[*] Corresponding author and topic chair: pia.thorngren@tsl.uu.se

CP628, *Women in Physics: The IUPAP International Conference on Women in Physics,* edited by B. K. Hartline and D. Li

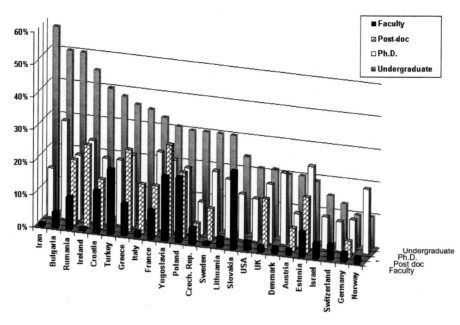

Figure 1. Currently estimated participation of women in physics at different levels of academia.

The reason for the north-to-south increase in Europe was discussed in one subgroup and found to be linked to, among other things, the fact that teaching is a highly valued profession for women in the southern countries, and female physics students there tend to go into teaching after graduating. This seems to be true for many African countries as well, where there is a lack of opportunities outside the educational system, and research is not a government priority. The conflict between the immediate needs of survival of large portions of the people and state funds for a traditional university is also a burning issue in countries like Brazil.

In the former communist countries, women physicists are found in greater numbers. In Bulgaria, for example, there are 50% women at the undergraduate level. Even though there is a "leaky pipeline" in these countries as well, the percentages of women physicists at higher levels are still high compared with those of the western hemisphere. Women were encouraged to go into science and to have a professional life the same as men under communist ruling, but the inequity persisted at higher ranks of the academia. Old structures of male dominance are not so easily changed, which was described in a very moving and honest plenary talk given by Iya Pavlovna Ipatova from Russia.

Furthermore, there is nowadays a "brain drain" of male physicists to the West, which leaves women physicists in greater relative numbers in the less developed European countries. Eastern European research often suffers from low financing, as well. In Latvia, for example, state support for science is so low that the paramount problem is a severe lack of funds and equipment; even computers are scarce. However, the governments of the other two Baltic countries seem to have recognized the value of research and university education and have a larger contribution from women scientists. It is worth noting that Estonia retains one-third of its women Ph.D.'s at the faculty level.

A system with entrance tests for admission to the university and a mandatory advanced level of mathematics and science in high school were pointed out as two reasons for the competitiveness of female students in some countries, such as Turkey, where women physicists make up 20% at both the Ph.D. and faculty levels.

In Iran and India the percentage of women in physics is high. Physics is not viewed as being a male subject and no barriers prevent women from entering the field. The relatively low prestige of the profession, which also seems to be true in Italy and some other countries, is a somewhat depressing reason for the proportion of women to be high.

Common Threads for Success/Failure

What are the most important positive and negative influences in different regions for women in sustaining a successful career? A few important common traits were found. The main positive influences for countries with a relatively high percentage of women in permanent positions are, according this discussion:

1. *Strong state social support* (e.g., well functioning child care).
2. *Supportive families* (husbands, mothers, etc.). Also named as an important factor by many respondents to the survey conducted before the meeting by the American Institute of Physics.
3. *Flexible working hours* (can be detrimental if one is expected to stay at work until late, as in Japan).

Among the most negative factors common to all countries—but with a stronger impact where few women physicists hold permanent positions—were the following:

1. *Age discrimination.* Given that women give birth and are mainly responsible for child rearing, age limits on grants and other opportunities work against women as combined age/gender discrimination. In developed countries some women, due to a lack of role models, do not discover the potentials of a physics career until later in life, resulting in an even more delayed career path.

2. *Nontransparent hiring process.* The rules for selection and promotion are neither well defined nor known to all. Jobs in academia are often inherited in an "Old Boys' Network" fashion. This is most detrimental when permanent positions are being filled.

3. *Too few opportunities for travel.* Mainly a problem for developing countries and the former communist countries. Membership in the European Union does not resolve this issue, because age limits are generally attached to traveling grants. As one conference attendee from Latvia put it: "Before we could not travel because of the iron curtain, now we are hindered by lack of travel money and because we are too old!"

4. *Late appointment for permanent positions.* The child-rearing period coincides with the period in life when it is expected of a dedicated physicist to go abroad for one or more postdoctoral assignments. In some developed countries the social security system works against women pursuing a scientific career. The academic ladder in Sweden, for example, contains another four- to five-year temporary position after the ordinary postdoctoral period. Using Sweden as an example: as in many countries, the cost of living in Sweden requires two incomes. Parental leave is 13 months with pay. If you are employed you get 80% of your previous income up to a certain limit, otherwise you get $12 per day. The difference can be as much as $19,000, which does not favor having an insecure job situation. Another barrier in Sweden is that if you go abroad you are without social security benefits, such as unemployment compensation and sick leave.

5. *Open sexual harassment* was reported to be a problem in two countries. It is a very serious issue that must be dealt with directly and promptly by the academic leadership.

RECOMMENDATIONS

The following recommendations were drawn from the discussions of the four subgroups:

- *Gather gender-specified statistics from all countries.* These efforts should be coordinated internationally, preferably by the physical societies and IUPAP in collaboration. Differences in education systems and universities make comparisons between countries extremely difficult. However, if data can be collected in a consistent way over time the effect of policy or other changes can be assessed and learned from. Make it mandatory for all universities to report gender-disaggregated statistics annually. Take care in defining professional level and use age as one independent variable (for an example from Japan, see Masako Bando's paper in these Proceedings). Clarify what local variations are valid for university entrance requirements and precollege education systems. Find dropouts and their reasons for leaving physics.

- *Implement transparent hiring processes and a fair evaluation system.* Define measures for promotion. Advertise positions openly, where women physicists will see them and become aware of the opportunity.

- *Create fellowships for returning to work, and fellowships with negotiable affiliation.* The Daphne Jackson Returners Fellowships in the UK are held only by those returning after a career break for family commitments. The pay is for half-time research for two years aimed at preparing the fellow for full-time work in either industry or university. Advanced Research Fellows in the UK are held by excellent young scientists of either gender for about five years and can be transferred to any university of choice.

- *Abolish age discrimination—use academic age.* The notion of academic age means that only the number of years after the Ph.D. exam are considered when assessing a candidate's merits. The clock must be paused during absences from the job market due to child rearing. Age limits create extra barriers for women to overcome. The Marie Curie Fellowship, for example, has an upper age limit of 35. Society misses a pool of excellent applicants this way.

- *Create worldwide exchange programs.* Provide fellowships for travel, conferences, Ph.D. programs, postdoc positions, and sabbaticals. Provide an internet forum for donations of used equipment and computers.

- *Network and establish Women in Physics groups; interact on a global scale.* This is beginning as a direct result of this conference.

Topic 6: Balancing Family and Career

Barbara Sandow[1*], Monika Bessenrodt-Weberpals[2],
Corinna Kausch[3], and Janis McKenna[4]

[1]*Freie Universität, Germany;* [2]*MPI für Plasmaphysik, Germany;*
[3]*Gesellschaft für Schwerionenforschung, Germany;*
[4]*University of British Columbia, Canada*

Abstract. The considerable challenges involved in balancing family and career have been identified as critical factors in the underrepresentation of women in physics worldwide. We review what were seen in discussion groups at the International Conference on Women in Physics to be the dominant issues for balancing family and career. We summarize the discussions and possible solutions (when possible solutions were found).

ISSUES AND POSSIBLE SOLUTIONS

Solutions and recommendations for helping to balance family and career would benefit all physicists—men as well as women. Thus, much of what follows is intended to apply to both sexes. We encourage men to play a role similar to that of women in terms of family responsibilities.

Travel and foreign work experience: Foreign work experience is often seen as beneficial, particularly in developing nations and small European countries, where it is often deemed essential for career progression. The inability to relocate, especially at the earlier graduate school or postdoctoral stages, can have a huge impact on the career of a young scientist. Often a spouse cannot give up a good job temporarily. The prospect of moving a family, especially when a spouse's job and/or children are involved, often precludes young physicists from gaining foreign work experience and hinders career progression. We recommend that physicists make plans early in their careers to accommodate foreign work experience, especially when they are from a country in which foreign experience is crucial for a successful career. Young physicists should try to stay flexible in the early stages of their careers.

Dual-career couples: A recent survey[1] conducted in the U.S. disclosed that 43% of married female physicists are married to other physicists (versus 6% of male physicists), and that over 68% of married female physicists are married to scientists (compared with 17% of male physicists). In Germany, a similar survey revealed that 65% of female physicists have a spouse with scientific academic training.[2] The difficulty in finding two professional jobs in the same geographic region is often referred to as the "dual-career problem." Often, the older partner finds a good job, and the younger partner trails along and may be underemployed. A number of North American and European institutes and universities now offer the possibility of split or shared hires (in which two people share 1 or 1.5 jobs), have spousal hiring programs in place, or help find alternative academic or nonacademic positions using headhunters or relocation specialists. Such programs should be widely encouraged and endorsed. A number of couples do end up maintaining two households and commuting long distances in order to secure two professional jobs. This last solution can work well for a limited time when there are no children involved. Some universities still have anti-nepotism rules that prevent a person from being hired if his/her spouse is already employed at the same institute or university. Such rules are outdated and we encourage institutes that have them to consider reexamining their policy.

Children: One of the greatest challenges in balancing career and family occurs when children arrive. For women, the crucial years for launching and establishing a physics career and gaining tenure often coincide with peak child-bearing years (the mid 20s to late 30s). Just when women need to be at the most productive stages in their careers, they often find themselves taking maternity leave and have large demands on time outside work for family reasons. To

---

* Topic chairperson and corresponding author: Sandow@physik.fu-berlin.de

CP628, *Women in Physics: The IUPAP International Conference on Women in Physics,* edited by B. K. Hartline and D. Li

some extent, men face a problem with this issue too, although often less severely because of the greater responsibilities usually taken up by the mother when it comes to family, and especially infant-care, issues.

Family leave: Many participants agreed that greater availability of maternity and paternity leave, family leave, and elder-care leave policies in industry and at universities and institutes would help the situation. Family leave policies instigated at the government level, regardless of workplace benefits, often make dealing with career and children easier. Participants noted that several countries have government-sponsored family leave acts, which in many cases even includes modest financial support during the leave. It was noted that in several countries in which paternity leave is available, very few male physicists take advantage of such leave. It is often perceived as stigmatized by society. Men should be encouraged to take paternity leave without shame and take pride in sharing family responsibilities.

Another issue was that of maternity leave for postdocs and graduate students. Postdoctoral and graduate student positions tend to be temporary soft-money positions with few or no benefits. Even when postdocs and graduate students have access to some type of maternity or paternity leave, they often cannot afford to take it when no income is associated with the leave. Universities, institutes, and granting agencies (which usually pay postdoc and graduate student stipends) should be encouraged to think seriously about establishing family leave policies with some modest funding for graduate students and postdocs associated with the leave. Also, fixed-term appointments should be extended by the period of the leave.

Participants noted that institutions should be encouraged to use a "career clock" for tracking career progress. The clock is stopped for family leave, as it is in many countries for military service, and slowed down for part-time work. Age limitations (when they exist) for scholarships, fellowships, and grants should use the career clock instead of age. Some universities already have policies in place to add time to the tenure clock (as well as to the sabbatical leave clock) if parental leave is taken; such policies should be widely encouraged.

Some women said they felt they had lost some respect or professional credibility because they took time off for maternity leave. (A serious physicist would not choose to stay at home with a child when she could continue in her quest to push the frontiers of basic knowledge, now, would she??) It might help change this point of view to publicize examples (and there are many) of women and men who take time off for family reasons then resume extremely productive and successful careers.

Flexibility: Job-sharing, a flexible workplace (e.g., telecommuting from home certain days of the week), flexible work hours, and reduced or part-time tenure-track positions should all be considered possible alternatives when compared with the prospects of losing women from physics careers entirely. Long-term part-time positions would probably not fly in an academic environment, but there have been cases in which women with full-time positions revert to part-time positions for short periods of time while children are young, and then return to full-time positions again once the children are a little older. Industry, institutes, and universities should consider allowing such flexibility. Also, the possibility of having some types of permanent positions available earlier in the career so that interruptions for family reasons might occur after the career is established should be considered.

Child care and family care: Child care is an issue that is handled very differently from country to country, but a common theme in developing and developed nations is that women tend to be responsible for the bulk of child, family, and elderly-parent care. The availability of high-quality child care for children of all ages, for the whole workday, and near the workplace was viewed as a high priority. This is presumably a high priority for male physicists, as well.

Child-care situations vary worldwide. A child-care solution for many women physicists in developing nations is to live near extended family, with grandparents or other relatives providing quality daycare. Whereas women from developing nations expressed concerns that the extended family puts many demands on women, it clearly also has some very positive benefits.

The age at which public school starts varies around the world. Public schooling begins at the age of 3 in some countries, and as late as 7 in others. Thus, in some countries child care is only a big issue for infants and toddlers, for a very short time while the child is very young, while in other countries child care must be secured for many years.

Another issue expressed with respect to child care was how it can be uncomfortable to pay someone to do a job that we find very important—raising our own children—rather than doing this job ourselves. In most countries, children's caregivers are paid low wages to do a job the mother would be doing if she were working at home rather than in physics. Hiring someone to be a substitute mother so that we instead do physics leaves many women with guilty feelings. Having child care at or near the work site would go some way toward alleviating this situation, because the parents could easily drop in on the child during lunch breaks or when they are concerned.

Extended families and elder care: Especially in developing nations, there are social and cultural expectations of women to play dominant roles and bear much responsibility in extended families. Thus, it is not easy for a woman to hold a serious scientific research position with large commitments expected for family needs. Some developing nations' participants said that their societies look unfavorably upon women with small children who return to work before the child is of school age. A woman with a professional career does not always gain the approval of the rest of her society; it will probably be at least another generation before such societies look at working mothers differently.

RECOMMENDATIONS

Our discussions yielded the following recommendations:
- Men should share in family duties as well as family pleasures.
- Governments should be encouraged to consider family leave policies (maternity and paternity leave) that have some financial support for the duration of the leave.
- Institutes should be urged to consider flexible work solutions, including shared or split jobs, the possibility of temporarily reducing a full-time position to a reduced-time position, and options for flexible work hours and telecommuting, when possible.
- Employers should consider initiating spousal hire programs and spousal relocation programs if they have not already done so.
- Any anti-nepotism rules still in place should be reexamined.
- Employers should be encouraged to support setting up childcare facilities at or near the workplace.
- Careers should be planned to include your private life.
- Physicists should try to be flexible in the early stages of their careers (you might have considerably less flexibility later on).
- Older people should mentor and support younger people and provide living examples of people who have successfully overcome the problems of balancing family and career.
- Many physicists who are working mothers have been pleasantly surprised that although combining a family and a physics career is challenging, it often is not as difficult as they had expected—women are encouraged to consider combining family and career.
- Make it clear to young physicists that doing physics for a career *does not* preclude having time left for a family life.
- Get organized and be confident!

CONCLUSIONS

Physics has never been a 9 to 5 profession. But physicists do tend to have families, and the quality, not quantity, of hours spent pursuing scientific problems should be stressed to young physicists. Many young people with concerns about balancing career and family simply decide to leave physics, because the perception is that in order to "make it," gain permanence, or get tenure, one must put in 70-hour work weeks. Many young men and women consider this a good reason *not* to stay in physics. We must dispel this notion and stress that it is the quality of work done, not the number of hours slaving away, that is most important. The image of the scientist should change from that of tousle-haired man who devotes every waking hour to his research, to the man or woman who combines achievement in an exciting job with a fulfilling family life.[3]

We did learn that people and families are all different—what works well for some people won't work for others. But the conflicts of a professional career in physics and family must be resolved, if we are ever to increase the participation of women in physics.

REFERENCES

1. L.E. McNeil and M. Sher, "Dual-Science Career Couples: Survey Results," http://physics.wm.edu/dualcareer.html; and L.E. McNeil and M. Sher, "The dual-career couple problem," Physics Today (July 1999).
2. B. Könekamp, B. Krais, M. Erlemann, and C. Kausch: "Chancengleichheit für Männer und Frauen in der Physik?" Physik Journal *1,* 22-27 (2002).
3. M. Osborn, T. Rees, M. Bosch, C. Hermann, J. Hilden, A. McLaren, R. Palomba, L. Peltonen, C. Vela, D. Weis, A. Wold, J. Mason, and C. Wenneras, "Science Policies in the European Union—Promoting Excellence Through Mainstreaming Gender Equality," report from the ETAN (European Technology Assessment Network) Expert Working Group on Women and Science (Office for Official Publications of the European Communities, Luxembourg, 2000); http://www.cordis.lu/improving/women/documents.htm.

Conference Recommendations

Many specific recommendations emerged during the conference. Not all will be applicable to all countries or situations. They should be reviewed by each country team, which should translate the applicable ones and work to implement them in their country. Other readers are urged to translate and work to implement the applicable ones, also. The recommendations are grouped into categories, but many are likely to have impact in other categories, too. When implemented, these recommendations will improve physics for both men and women.

GENERAL RECOMMENDATIONS

1. Coordinate internationally the collection and access of data on physics demographics, including gender, to watch and influence trends. Collect data regularly (every one to three years) and in a consistent way, to watch and influence trends. Request data from national and regional physical societies. Find out, also, why women leave physics.

2. Create, support, and encourage networks for women physicists: local, national, and international, including a worldwide e-network. Create women-in-physics web pages in each country, with links to each other and to information on successful strategies and programs. Provide a well publicized international web presence for women in physics.

3. Involve men, especially highly respected physics leaders, in improving the climate for women (and minorities) in physics.

4. Have transparent, gender-blind processes for important decision making. Transparency can be aided by requiring decisions to be reported and explained. Important decisions include those related to recruitment, selection, salary, promotion, peer review, conference programs, allocation of space and equipment, and other issues affecting important working conditions.

5. Establish mechanisms for assessing and improving the climate for women (and minorities) in physics. Proven approaches include creating special committees for women in physics and focusing resources and attention on this issue. For example, a government or university could provide funding (or partial funding) for the initial years of a tenure-track position if it is filled by a woman. Special committees could visit universities, research institutes, and other employers of physicists to advise on their climate for women.

6. Encourage written rules and policies (for example, an equality policy) to achieve fairness and transparency in policies, practices, and decision making.

7. Create index of web links to international funding sources.

8. Remove barriers to full participation of girls and women (for example, insufficient restrooms and dormitories).

9. Adjust the reward structure at all levels to encourage desired behaviors.

ATTRACTING GIRLS INTO PHYSICS (CHILDHOOD TO UNIVERSITY)

1. Revise educational curricula and materials to connect physics with medicine, biology, technology, the environment, etc., and to show diverse physics career paths and job prospects. Ensure that physics courses, math courses, textbooks, equipment, and funding for girls' education are as good as for boys' education, and feature women physicists as role models.

2. Strengthen the training of science/physics teachers and include opportunities for them to do research and to interact with working scientists. Train teachers and counselors in gender issues (girl-friendly classroom

CP628, *Women in Physics: The IUPAP International Conference on Women in Physics,* edited by B. K. Hartline and D. Li
© 2002 American Institute of Physics 0-7354-0074-1/02/$19.00

atmosphere, examples of interest to girls). Attract qualified school teachers with fair pay, respect, and good working conditions.

3. Publicize physics role models who counteract the stereotypes and whose stories are examples of career success and leadership positions.

4. Educate parents about opportunities for daughters and how to encourage them.

5. Encourage "smart girls networks" (clubs, enrichment opportunities, etc.).

6. Attract more girls to compete in prestigious physics competitions.

7. Raise boys to share family responsibilities and to expect women to have professions.

8. Get international help and funding for schools in developing countries

9. Involve universities, research institutes, and industries in helping schools and strengthening teacher training.

LAUNCHING A SUCCESSFUL CAREER (UNIVERSITY TO MID-CAREER)

1. Have flexible entry and graduation requirements for physics majors, and provide early opportunities for students to participate in research.

2. Train/sensitize faculty and supervisors to gender issues (female-friendly atmosphere, respectful and collegial treatment).

3. Provide enlightened and supportive mentors and supervisors for women physicists. These persons should find funding, teach women the "rules of the game" and how to write successful proposals, introduce them to important professional contacts, give them challenging assignments and opportunities, provide constructive feedback on unsuccessful proposals or interviews, give them credit, and advocate them in the physics community.

4. Provide training for women physicists in presentation of results, paper writing, grant applications, etc.

5. Shorten the post-postdoc phase with its inherent insecurity and relocation requirements.

BALANCING FAMILY AND CAREER

1. Respect and value family obligations (quality child care convenient to workplace and at conferences, flexible working hours).

2. Pause "career clock" and have flexible age limits and rules for grants and fellowships, so as not to disadvantage people who take time for family responsibilities. (Accord career interruptions for "family service" the same respect as for military service.)

3. Provide funding sources to help people return to physics after a career pause.

4. Solve the dual-career couple problem by facilitating geographically co-located job opportunities and creative solutions, such as shared positions.

GETTING WOMEN INTO PHYSICS LEADERSHIP

1. Appoint women physicists to leadership positions and include them on important committees in their institutions, countries, and professional societies, and in IUPAP.

2. Involve more people in leadership. Consider innovative approaches, such as shared positions, term appointments, and novel structures.

INTERNATIONAL ASPECTS

1. Create opportunities for R&D employment, funding, and research equipment in developing countries.

2. Provide opportunities for collaboration and exchanges between regions and countries. Provide resources for conference travel for physicists from developing countries, and for physicists from developed countries to be visiting lecturers in developing countries.

3. Establish and sponsor international speaker program(s) for women physicists that would include web-accessible database of names and topics, and travel support.

4. Sponsor prestigious, topical international physics summer schools with female and male speakers, organizers, and participants.

PART TWO

Conference Proceedings

Opening Remarks

A Multiplicity of Perspectives:
With Understanding Can Come Improvement

Burton Richter

President, International Union of Pure and Applied Physics

Distinguished Guests, Delegates, Ladies and Gentlemen, welcome to this International Conference on Women in Physics. Bringing this conference to realization has been one of the main efforts of IUPAP since our last General Assembly in 1999. This meeting may well be unique. While there have been many studies in the last few years of the roles and problems of women in science that are local, national or regional, this is the first to involve so many nations. You come from 65 countries with different cultures, and have different backgrounds and different perceptions of the issues. We very much hope that by bringing you together, your multiplicity of perspectives will generate a new and broader view of the problems faced by women in science, and generate an action agenda that each group can take home and that international organizations might adopt.

There are two issues that I hope you can address. One question is why the fraction of women in physics is so small, and how that fraction might be increased. The other question, probably more important and certainly related, is what might be done to improve the situation of women in physics so that they share appropriately in jobs, professorships, and positions in the power structure.

There is obviously no intellectual impediment to a successful career by a woman in physics. However, there seem to be many societal ones. There are the obvious: the schooling of girls, family responsibilities, lack of effective and affordable child care, a male-dominated world view that will yield only slowly to relentless pressure. Change will come more quickly in societies that have already recognized that there is a problem than it will in societies where tradition makes the role of women such that the culture sees no problem at all. However, I believe that working together and maintaining contacts among all of the groups present here will help to speed change in all societies.

There are, however, issues beyond the obvious. In the United States 40 years ago only 2% of the physics Ph.D.'s went to women. This number began to rise with the beginning the women's movement and reached about 13% a decade ago, where it has stayed ever since. Meanwhile, in biology the number of Ph.D.'s continued to rise and has reached 50%. In chemistry it has reached 30%. Only mathematics and engineering have a lower percentage of women Ph.D.'s than does physics. Why is this, and what might we do about it?

In Europe, Scandinavia has been the earliest group of countries to eliminate barriers to women in all fields, including politics, industry, and education. Yet Scandinavia has the smallest number of women in physics in all of Europe.

Pakistan can have a woman as Prime Minister, but almost never has a woman as professor.

I hope the discussions among all of you with your different perspectives can shed some light on this problem. With understanding can come improvement.

IUPAP's Working Group on Women in Physics has assembled you here in the hope that your different experiences and the different problems you have faced and overcome can produce ideas on how to address the obvious problems, shed some light on the not-so-obvious problems, and generate a broad agenda for change. What comes from your discussions can have an influence beyond physics. Other scientific societies have supported our efforts to bring you here and are looking to our results to give clues to what they might do. We hope that the multicultural approach can generate something new.

I look forward to listening, and I will do my best to implement your recommendations.

Thank you and welcome.

No Less Than the Future of Our Planet Is at Stake

Walter Erdelen

Assistant Director-General for Natural Sciences, UNESCO

Commissioner of the European Union, Mr. President of the International Union of Pure and Applied Physics, Ladies and Gentlemen, on behalf of the Director-General of UNESCO, Mr. Koichiro Matsuura, I should first like to welcome you all to UNESCO Headquarters. In doing so, let me say what a pleasure it is for me to address, in this building where so many international meetings have taken place over the years, the first International Conference on Women in Physics.

It is of particular satisfaction to greet so many women scientists on this, the eve of International Women's Day. Satisfaction, also, in welcoming so many women scientists who have found success in a domain that is, alas, still too often considered a male reserve—physics research.

Finally, satisfaction in helping to host this event with the International Union of Pure and Applied Physics, a family member of an organization with whom we have enjoyed such good and effective relations over the years—the International Council for Science (ICSU).

Our partnership with ICSU was marked by the joint organization of an event that was at the very heart of why we are gathered here in Paris today: the World Conference on Science, a meeting that established the basis of an alliance between science and society for the 21st century. From June 26 to July 1, 1999, in Budapest, Hungary, 1800 delegates including representatives of 159 countries, 79 international nongovernmental organizations, and 79 academies of science and research organizations came together to discuss how best to organize science in such a way as to respond to the needs and aspirations of society, and to provide it with the means to do so.

In essence, the Budapest Conference allowed us to set out the terms of a new social contract between science and society. Of course, the putting into place of a new commitment by all, toward a new science for all, implies a much more democratic decision-making process than hitherto—in other words, the active participation of a greater number of people, and particularly women, so long separated for cultural, religious, and political reasons from a field of activities assigned to men.

I can tell you that we shall be eagerly awaiting the results of your conference in order to develop our analysis of the problem. Anecdotal accounts and available statistics all serve to show that in all countries—to one degree or another—the involvement of women in the progress of science is very much lower than that of men, especially in certain disciplines such as physics. Another more telling fact, in whatever field of research you care to choose, is the virtual absence of women from positions of influence or decision making.

This waste of human resources, this injustice for so many human beings, this intellectual impoverishment of research, must be unreservedly denounced and brought to an end.

Thanks to the unprecedented mobilization of women that we saw in Budapest, and the unwavering support of many male delegates, important recommendations were adopted to this end: access to scientific education, development of careers, science policies, networking, to mention just a few. The holding of this conference today is one of the more tangible signs that the dynamism generated in Budapest has borne fruit.

Ladies and Gentlemen, the question of the role of science and the responsibilities of scientists will soon be once more on the international agenda. Ten years after Rio, a World Summit on Sustainable Development is being convened in Johannesburg by the United Nations. A first statement has just been made by the Secretary-General of the summit, who has announced in his report that "progress achieved in meeting the objectives of Rio has been slower than foreseen and, in certain cases, the situation is even worse than 10 years ago."

Faced with the emergencies with which society is confronted—eradication of poverty; coping with changes in production and consumption patterns, especially in the field of energy; management of natural resources and the protection of the environment—to mention only the most pressing, what will the response of scientists be? On questions as fundamental as the management of freshwater, the monitoring of climatic change, research into sources of energy that are nonpolluting and renewable, the mitigation of natural disasters, among others, how

CP628, Women in Physics: The IUPAP International Conference on Women in Physics, edited by B. K. Hartline and D. Li
© 2002 American Institute of Physics 0-7354-0074-1/02/$19.00

43

are scientists going to position themselves? In these days when many of the major research projects are multidisciplinary, and in which physics represents a transversal component, what is to be the reaction of physicists?

In the debate around the ethics of science, which will extend far beyond Johannesburg, the women in physics will need to make their views known. They will have to express their needs publicly, propose their solutions, and be involved individually or collectively, whatever their level of responsibility.

Will they do this in a different way from their male colleagues? To what measure will they adopt new positions? We shall know better when they enter in greater number into the scientific field and, above all, when they attain equality with their male colleagues at decision-making levels.

But one thing is certain: their voice—your voice—will be indispensable for the democratic debate on no less a subject than the good governance of the planet and the very survival of the human race.

To my mind, this Women in Physics Conference opens the road to all possibilities.

Commissioner, Excellences, fellow scientists, the conclusions and recommendations that you will formulate here in Paris will be a first but decisive contribution of the community of physicists—men and women—toward the collective management of research, and more largely, of our world as we strive for economically effective yet socially equitable and ecologically sound development.

To all of you I say *merci et bravo* and wish you an enjoyable and effective three days in UNESCO.

Thank you.

A New Momentum for Gender Equality in Research

Philippe Busquin

Commissioner for Research, European Union

I am very pleased to have the opportunity to address you at the opening of this conference, the first to be dedicated to the promotion of women in physics. I am in fact a physicist by training myself and I can, therefore, appreciate the extent to which the challenge you have set yourselves is both difficult and necessary.

ACTIONS BY THE EUROPEAN COMMISSION

Under the heading "Women and Science," the European Commission is currently engaged, through a variety of actions, in the promotion of more equality between men and women in scientific research. With these actions, the Commission aims to act as a catalyst to promote the participation of women in scientific research. I will let Professors Teresa Rees and Claudine Hermann present these activities: they have played leading roles in the new momentum for gender equality in European research.

Equality between men and women in research is a basic condition for the achievement of the European Research Area and, to take this one step further, for a knowledge-based society. When full and formal equality between men and women is declared and guaranteed and when university degrees are equally divided between men and women, it will be unthinkable that scientific research continues to be an area where women are only tolerated.

A FEW HISTORIC EXAMPLES

The question of whether science would be different if there were more women is a question we have to live with but one to which we can never reply.

The example of Agnes Pockels, a pioneering physicist of the 19th century, is quite revealing. At the time, scientists were studying how to measure the surface tension of water. As you know, clean surfaces are a major problem for surface experimentation but Agnes Pockels found an original and simple solution. She sent a polite letter to Lord Rayleigh explaining her method and he, the famous physicist, realizing the excellence of her work, sent the letter to the journal *Nature*—and it was printed. The technique she developed became standard procedure.

Look too at Marie Curie. She would not have received the Nobel Prize for her discoveries if Pierre Curie, who learned he was about to receive the prize, had not made it known that he wanted Marie to be associated.

Lise Meitner, as you know, had not been associated with the Nobel Prize for Chemistry given to her collaborator Otto Hahn in 1946. The two women who won the Nobel Prize for Physics (Marie Curie in 1903 and Maria Goeppert-Mayer in 1963) on each occasion received it along with two other people.

Exceptional examples such as these demonstrate that the gender question interferes with the recognition of scientific excellence. And we know in what difficult circumstances Marie Curie received her second Nobel Prize, in a storm provoked by her amorous relationship with Paul Langevin. Never was it considered pertinent to question the scientific excellence of a man, in similar circumstances.

NEED FOR POLITICAL DEBATE

These biographical examples also suggest that nothing is self-evident in terms of equality between men and women in scientific research. It is our responsibility—all of us, political and scientific decisionmakers—to work

CP628, *Women in Physics: The IUPAP International Conference on Women in Physics,* edited by B. K. Hartline and D. Li
© 2002 American Institute of Physics 0-7354-0074-1/02/$19.00

together with the aim of making scientific research more welcoming for the talents of women. I will ensure that the European Commission will play its role in the areas where it has competence.

I thank you all very much indeed and wish you every success in your research and in your determination to push further for gender equality in research, and particularly in physics.

International Survey Report

Women Physicists Speak:
The 2001 International Study of Women in Physics

Rachel Ivie, Ph.D., Roman Czujko, and Katie Stowe

Statistical Research Center, American Institute of Physics

ABSTRACT

Abstract. The Working Group on Women in Physics of the International Union of Pure and Applied Physics (IUPAP) subcontracted with the Statistical Research Center of the American Institute of Physics (AIP) to conduct an international study on women in physics. This study had two parts. First, we conducted a benchmarking study to identify reliable sources and collect data on the representation of women in physics in as many IUPAP member countries as possible. Second, we conducted an international survey of individual women physicists. The survey addressed issues related to both education and employment. On the education side, we asked about experiences and critical incidents from secondary school through the highest degree earned. On the employment side, we asked about how the respondents' careers had evolved and their self-assessment of how well their careers had progressed. In addition, the questionnaire also addressed issues that cut across education and employment, such as the impact of marriage and children, the factors that contributed the most toward the success they had achieved to date, and suggestions for what could be done to improve the situation of women physicists.

HIGHLIGHTS

- The report contains country-level data and anecdotal information about the representation of women in physics from 34 countries.

- Most of the women physicists who responded to this survey reported that they developed an interest in physics during or before they were in secondary school. This emphasizes the importance of the opportunity to study physics and the encouragement to pursue science early in the academic system.

- Respondents felt they had generally positive experiences as undergraduates and as graduate students.

- About one-third of the women who responded felt that they had progressed more slowly in their careers than their colleagues had.

- The demands of a career in physics seemed to preclude several of the women in our study from marrying or having children. Of those who are married, a significant number reported that marriage affected their work. When comparing themselves to their colleagues, women with children were more likely than women who do not have children to say their careers had progressed slowly.

- The factor most frequently cited by women physicists as contributing to their success was the support of their families, including their parents and husbands. Many also mentioned the support of advisors, professors, and teachers, and some cited the support of colleagues. Also frequently mentioned were the women's own determination, will power, and hard work.

- Barriers that the women mentioned included the problems of balancing the demands of child care with the demands of a scientific career. Another barrier was discriminatory attitudes, usually expressed in the form of assumptions that women cannot do physics.

- Three out of four women who responded said that they would choose physics again.

CP628, *Women in Physics: The IUPAP International Conference on Women in Physics,* edited by B. K. Hartline and D. Li
© 2002 American Institute of Physics 0-7354-0074-1/02/$19.00

SECTION I: COUNTRY-LEVEL DATA AND ANECDOTAL INFORMATION

This section contains country-level data and anecdotal information on the representation of women in physics from various countries.

Country-Level Data

The country-level data (Table 1) were compiled from external, reliable sources such as government agencies and physics societies that routinely collect data such as these. In order to be included in the country-level data, countries had to provide accurate numbers of physics graduates broken down by year, level of degree, and gender.

TABLE 1. Percentages of Physics Degrees Awarded to Women in Selected Countries, 1997 and 1998 (2-year averages).

Country	Ph.D.'s %	First-Level %
France	27	33
Poland[a]	23	36
Norway[b]	23	20
Ukraine[c]	23	—
Australia[d]	22	20
Turkey	21	37
India[e]	20	32
Columbia[f]	—	28
Denmark	17	19
Lithuania[g]	17	—
United Kingdom	16	20
China-Taipei	13	19
United States	13	18
Sweden	13	17
Canada	12	22
Mexico[h]	10	18
Germany[i]	9	10
Switzerland[j]	9	9
The Netherlands	9	5
South Korea	8	30
Japan	8	13

[a] Poland: 1998 data only.
[b] Norway: 1996-2001 data.
[c] Ukraine: 2000-2001 data.
[d] Australia: Ph.D. data include some master's degrees (higher degree by research).
[e] India: Partial data from the Registrar General of India, 1998.
[f] Columbia: 1998 and 1999 data.
[g] Lithuania: 1996-2001 data.
[h] Mexico: Ph.D. data for 1998 only. Bachelor's data for 1998 and 1999.
[i] Germany: Includes astronomy and astrophysics.
[j] Switzerland: 1999-2000 data.

Anecdotal Information

The anecdotal information was collected mainly through personal correspondences of the IUPAP team leaders with universities and colleges in their countries. Although some team leaders were able to obtain fairly complete information, it was not appropriately broken down by year, and so could not be included in the comparison of country-level data. A complete list of sources is at the end of the report.

Albania
First-Level Degrees
- The three institutions that award first-level degrees for a teacher of physics reported awarding 16 degrees to women and 13 degrees to men during the 1999-2000 school year.

Armenia
First-Level Degrees
- The one institution that grants first-level degrees reported awarding 30 degrees to women and 35 to men during an unspecified time period.

Ph.D. Degrees
- The six institutions that award Ph.D. degrees reported awarding one degree to a woman and three to men during an unspecified time period.

Cameroon
First-Level Degrees
- The University of Dschang reported awarding 24 degrees to women and 131 to men since 1993.

Chile
First-Level Degrees
- The Pontificia Universidad Catolica de Chile reported awarding 10 degrees to women and 81 to men since 1978.
- The Universidad de Santiago de Chile reported awarding eight degrees to women and 55 to men from 1990 to 2000.

Ph.D. Degrees
- The six institutions that grant Ph.D.'s in physics reported awarding six degrees to women and 54 to men since 1975.

China-Beijing
- The Chinese Physical Society reported that 15% of its members are women.

Columbia
Ph.D. Degrees
- The four institutions that grant Ph.D.'s in physics reported awarding one to a woman and five to men during an unspecified time period.

Croatia
First-Level Degrees
- One of the four institutions that grant first-level degrees reported awarding approximately 10 degrees to women and 20 to men during an unspecified time period.

Ph.D. Degrees
- The two institutions that award Ph.D. degrees reported awarding approximately one degree to a woman and four to men during an unspecified time period.

Czech Republic
Ph.D. Degrees
- During an unspecified period of time, five Ph.D. degrees were awarded to women and 30 were awarded to men.

Estonia

First-Level Degrees

- Two of the three institutions that grant first-level degrees reported awarding 16 degrees to women and 60 to men between 1999 and 2001.

Ph.D. Degrees

- One of the two institutions that grant Ph.D. degrees reported awarding one degree to a woman and 12 to men between 1999 and 2001.

Iran

First-Level Degrees

- 46 of the approximately 66 institutions that grant first-level degrees reported awarding 200 degrees to women and 350 to men during an unspecified time period.

Ph.D. Degrees

- Nine of the 12 institutions that grant Ph.D. degrees reported awarding three degrees to women and 25 to men during an unspecified time period.

Israel

Ph.D. Degrees

- At Tel Aviv University there were 12 Ph.D. degrees awarded to women and 53 awarded to men from 1998 to 2001.

Lithuania

First-Level Degrees

- The five institutions that grant first-level degrees reported awarding 84 degrees to women and 184 to men during 2000 and 2001.

South Africa

First-Level Degrees

- The University of Potchefstroom reported awarding six first-level degrees to women and 61 to men over the last 10 years.

Ph.D. Degrees

- The University of Port Elizabeth reported awarding one Ph.D. to a woman and five to men between 1991 and 1995.
- The University of Potchefstroom reported awarding two Ph.D.'s to women and 12 to men over the last 10 years.
- The University of Cape Town reported awarding one Ph.D. to a woman and four to men over the last five years.

Sudan

First-Level Degrees

- Three of six institutions that grant first-level degrees reported awarding 56 degrees to women and 97 to men during an unspecified time period.

Federal Republic of Yugoslavia

First-Level Degrees

- Five of the six institutions that award first-level degrees reported awarding 561 degrees to women and 525 to men over the last 15 years.

Ph.D. Degrees

- The four institutions that grant Ph.D.'s reported awarding 76 degrees to women and 266 to men since 1947.

SECTION II: THE INTERNATIONAL SURVEY OF WOMEN IN PHYSICS

How the Survey Was Conducted

There is no source for the names and contact information for all physicists (irrespective of gender) in every country. Thus, we relied on personal networks among women physicists in each IUPAP country. The team leaders in each country were asked to e-mail women physicists they knew in their countries. The latter were asked to complete the questionnaire and return it via e-mail to the Statistical Research Center, as well as to forward the questionnaire to other women physicists they knew. During the summer and early fall of 2001, Dr. Marcia Barbosa, chair of the Working Group, e-mailed team leaders in IUPAP countries on several occasions to encourage them to disseminate the questionnaire. We eventually received more than 1000 responses from women physicists in more than 50 countries. Table 2 shows the number of responses by country.

TABLE 2. Number of Women Physicists who Responded to the International Study by Country, 2001.

Continent/Country	Number	Continent/Country	Number
Africa		**Europe**	
Cameroon	2	Albania	6
Egypt	13	Armenia	8
Nigeria	20	Austria	4
South Africa	23	Belarus	2
Tanzania	1	Belgium	26
Zimbabwe	2	Bulgaria	3
Asia		Croatia	13
China-Beijing	27	Czech Republic	5
China-Taipei	5	Denmark	17
India	42	Estonia	5
Indonesia	16	Finland	1
Israel	9	France	37
Japan	57	Germany	30
Malaysia	1	Greece	2
Pakistan	2	Ireland	72
South Korea	21	Italy	41
Turkey	73	Latvia	13
Uzbekistan	1	Lithuania	2
Australia/New Zealand		Netherlands	22
Australia	16	Norway	1
New Zealand	1	Poland	4
North America		Portugal	1
Canada	48	Romania	5
Cuba	6	Russia	24
Mexico	9	Spain	21
USA	82	Sweden	12
South America		Switzerland	17
Argentina	27	UK	47
Brazil	33	Ukraine	3
		Yugoslavia	20

The authors worked with the IUPAP Working Group on Women in Physics to develop several versions of the questionnaire instrument. In light of the unique character of the educational and economic systems in each country, we decided to focus on a few key issues and a few key stages in the education and careers of physicists. It was decided that to attempt a more detailed picture would have required a questionnaire of considerable length, which would have been an enormous burden on the respondents. Even so, the final questionnaire instrument grew to six pages in length.

The findings from the international survey include women physicists who responded even if they do not currently live in an IUPAP country. However, the report on this study does not include responses either from men physicists or from those women whose highest degrees were in fields other than physics.

Cautionary Comments

Readers of this report are cautioned about the generalizability of the data from the survey. First, although we heard from women physicists who currently reside in more than 50 countries, their experiences may not be representative of all women physicists in all countries. In fact, the personal networks that made it possible to contact women physicists may also bias the results in favor of those women who are known and reachable by e-mail. Thus, in some countries, the women who responded may overrepresent the number of women physicists who are academically employed. Also, e-mail was used almost exclusively in the distribution of the questionnaire. However, e-mail is not yet universally available, and so some women physicists may not have had the opportunity to participate in the study.

Second, the study only includes those women physicists who persisted. Students who dropped out of physics during their education could not be identified. Had we been able to include such students, we may have received more negative comments about the educational systems in different countries. We were unable to locate women physicists who left physics after earning an advanced degree. Thus, the survey did not have adequate representation from women physicists who could not find jobs in physics, those who left physics because they had children and were expected to devote their time to their families, or those who left physics because of discriminatory attitudes. Finally, the survey did not include women who did not have the opportunity to pursue advanced education because of the economic, social, political, or educational systems in their countries.

Finally, the questionnaire instrument was written in English, and some women physicists are not fluent in English. Thus, some questions may have been misunderstood. Also, we saw evidence in the replies to open-ended questions that some respondents were having difficulty expressing themselves using English.

Encouragement From Parents

Most of the women physicists who responded to this survey reported that they developed an interest in physics during or before they were in secondary school (Table 3). Thus, early encouragement to pursue education and an early exposure to science are fundamentally important, and family members are in an excellent position to provide these experiences.

Parents play an essential role in the development of their children. They can encourage their children to pursue education and to pursue science. They can provide their children with hands-on scientific experiences and can expose children to exciting scientific events going on around them. Parents also play an essential role in the development of a child's self-esteem. This strong belief in one's intellectual ability is a critical source of strength

TABLE 3. When Did You First Think of Choosing Physics as a Career?

	Percent[a]
Before high school	13
During high school	58
During first-level degree	24
During graduate degree	4

[a] Percents do not sum to 100 due to rounding.

during the rigors of physics education. Confidence in one's ability can be especially important for female students when they confront the negative effects of sexism, which can cause women to question their ability or their right to pursue advanced degrees.

"My parents always expected a lot from me and gave the required support. I learned very early to believe in myself. I have also met professors with a lot of prejudice against women, but as I had a sound self-esteem, it did not really harm me." [Brazil]

"My dad always encouraged me in mathematics and science and was never too busy to answer questions and foster my interests in science. He encouraged me to go to university, saying it would be a waste of a good brain if I did not go." [South Africa]

"Both my parents supported my interest in science. My parents bought me a telescope, a chemistry set, geology kits, and electronic kits. My mother woke me up for every US space launch and watched the paper for special astronomical events." [USA]

"I owe a great deal to my father. Reading books and biographies of scientists, books like 'Fun with Science' motivated me for research." [India]

"My family's support, particularly my mother's [was important to my success]. She is an elementary school teacher and always encouraged me to study." [Mexico]

"My father was a physics teacher and taught me the beauty of physics. I received an education focused on sciences and never felt I was treated different for being a woman." [Switzerland]

"My family allowed me to study as much as I wanted." [South Korea]

"I think I did it all by myself, but I can't ignore my big brother's support during my education." [Turkey]

Encouragement From High School Teachers

High school teachers have a critical role on several levels. They have the responsibility of teaching students about both the subject matter and the excitement of a field. They also have the opportunity to affect students' confidence in their ability to succeed during the period when many students first make choices about eventual fields of study.

Almost one-third said they were influenced to choose physics by their teachers. However, about one-fourth rated the teaching of high school physics as worse than other subjects in their country.

"I was very fortunate to have an excellent science teacher in high school. We did experiments and I was utterly fascinated by the accuracy in some of the results. I think that having the opportunity to actually see first-hand the confirmation of what one learned in class made a deep impression on me and was very satisfying." [South Africa]

"I had talent in physics and mathematics, but the middle school teacher of mathematics (he was a male) didn't care for me. He was dismissive of girls' abilities and didn't like for girls to be the first in the class. While the teacher of physics (she was female) took care of me, she supported me and encouraged me to study physics, worked with me by giving me difficult exercises, laboratory work and different books in physics." [Albania]

"My secondary school teacher was a scholar and taught us so well that everyone in my class was interested in physics." [Nigeria]

Undergraduate Education

Undergraduate education is a unique stage in the educational system for several reasons. It is the first time that students publicly declare their intention to pursue a particular field of study. It is the first time that many students will meet practitioners in their intended field. It is also the first time that one field will make up a considerable portion of their studies.

Professors play an essential role during this stage of education. Imparting knowledge about the discipline is but one aspect of their responsibilities. No two students enter college with identical educational experiences and expertise. The best professors typically spend time with each of their students to identify the student's strengths and weaknesses. They then advise their students on ways to build on their strengths and address their educational shortcomings. Finally, professors represent the front line of their discipline. Thus, it falls to them to explain the excitement of their field and to encourage students to go as far in the educational system as their abilities will take them.

TABLE 4. Number of Undergraduate Physics Majors and Quality of Attention From Professors.

| Number of Majors | Attention[a] | |
	Positive %	Neutral %
Fewer than 10	67	32
10–89	61	32
90 or more	45	45

[a] Rows do not sum to 100% because a few respondents chose "Negative" or "Other."

Almost all of our respondents were undergraduate physics majors, and almost all who were said they had received either the same amount of attention from their professors or more attention from their professors than the other students did. However, class size affects the quality of the attention that female physics students received (Table 4). Students who graduated from large departments reported less positive attention than students who graduated from smaller departments.

It should be noted that students who dropped out of physics were not included in this study. Thus, the fact that we received only a few negative comments about the attention professors gave to female students may not be representative.

"[When I was] an undergraduate, the majority of my professors and instructors were very good and very helpful. There was always a member of the department available to help with any questions or problems I had. The value of positive, enthusiastic teaching staff is enormous." [Canada]

"I got a lot of attention, all the help I needed—and sometimes more. I felt respected by the professors at the institute." [Denmark]

"I only decided to purse a career in physics during my undergraduate years, as a result of having some excellent and inspiring lecturers." [Ireland]

"My experience with university professors was patchy. Some were always ready to help, never too busy to answer a question and never made me feel small. Others were simply indifferent, while others took a special delight in taking students down a peg, reminding us that we were not as lofty as they and all but discouraging us from continuing with physics as a career. From these last groups I feel that I received little beyond the actual material presented to me by them." [South Africa]

Graduate Education

Graduate physics education provides students with knowledge of physics and an understanding of the basic principles that govern how the physical world works. However, a physics education does even more. It also provides students with the research and cognitive skills that are important for a good physicist. These include critical and analytical thinking, problem-solving skills, and learning how to define a problem and identify a set of possible solutions. In graduate school, students learn to look at the world in the unique way that physicists do.

"I believe that a...physics degree develops the ability to think laterally, to work through processes and consider all avenues before development." [Ireland]

In addition, physics students also acquire technical skills such as advanced mathematics, computer skills (both hardware and software), and the ability to work with lab equipment, including using, designing, building, and repairing sophisticated equipment. Finally, physics provides students with educational experiences that are intended to develop the traits that are important in good scientists, such as being hard working, meticulous, persistent, tenacious, and self-confident. In other words, in graduate school, students learn the culture of physics. They learn how to think like a physicist, communicate like a physicist, and work like a physicist.

Given the importance of graduate school, a good advisor can make all the difference in the professional success of a young physicist. The women who responded to our survey tended to have supportive graduate school advisors. Almost all of our respondents said that they had no difficulties finding an advisor, and almost all of the advisors were male. More than four out of five described their relationships with their advisors as either excellent or good, and most say that their advisor treated them better or the same as other graduate students.

"[I had a] female advisor/mentor in graduate school who made sure I was introduced into the scientific community, learned how to write proposals and papers, and otherwise did all the things advisors ought to do for their students but, in my opinion, rarely do." [USA]

"I was always pushed forward by my Ph.D. chiefs—two women!!!" [France]

However, graduate school was not a welcoming environment for all of our respondents.

"In graduate school, I was in an extremely hostile environment and miserable with no support from any of my research group. But, I was tough and fought to stay in grad school. It was an incredibly sexist place, but the fact that if I quit, that would cut the number of women in the physics class by 50%, kept me struggling to survive." [USA]

Graduate education is the time when students learn to do research. In a positive educational atmosphere, students present their research at conferences and learn to write professional papers. During graduate school, almost three-fourths of our respondents were asked to present a poster or a talk at a conference, and two-thirds were asked to co-author a paper. However, only about half were asked to write a research paper on their own while they were in graduate school (Table 5).

TABLE 5. Percentage of Responding Women Physicists Who, in Graduate School, Were Asked to:

Present a poster/talk at a conference	72%
Co-author a research paper	66%
Write a research paper independently	47%

Exposure to International Research Opportunities

A physics Ph.D. is a research degree, and opportunities to do physics research are essential. A significant aspect of developing research expertise involves participation in international research projects. Many physicists make research visits to foreign countries or even move internationally to pursue careers. However, most of our respondents went to undergraduate school and graduate school in the same country in which they currently work. Three-fourths of these have made research visits to foreign countries at some point in their careers.

"A specific opportunity [that led to my success] was the fact that my university was closely related to the Joint Nuclear Research Institute (JINR). I visited JINR 11 or 12 times for short stays of about one month, but it was very important to meet people and to work in an international environment." [Bulgaria]

"The support of my research advisor was integral to my success. Specifically, the opportunity to interact with international researchers and to attend and present my work at conferences was a major factor in attaining my current status." [Canada]

"While at university, I had the good fortune of being able to spend several months before my senior undergraduate year as a summer research intern at a university in the USA and then the summer before my graduate studies at the CERN summer school. This gave me invaluable insights into modern physics research and gave me contacts and interests that helped form my career in a decisive way." [Germany]

"I worked in a group of young theoreticians with a successful leader who had international contacts." [Estonia]

Career

The importance of a Ph.D. for becoming a professional physicist varies by country. For example, Italy did not offer Ph.D.'s in physics until the 1980s. Most of our respondents, however, do have Ph.D.'s (Table 6).

TABLE 6. Highest Degree Obtained by Women Physicists Who Responded.

	Percent
Ph.D. or higher	65
Less than Ph.D.	20
Current student	14
Unknown	1

In addition to Italy, there are a few other countries where many of our respondents do not have Ph.D.'s (Table 7). It may be that physicists in these countries work without Ph.D.'s. On the other hand, a low percentage of respondents with Ph.D.'s may be an artifact of the sample. Respondents knew each other, and someone without a Ph.D. may have forwarded the questionnaire to her colleagues who also did not have Ph.D.'s. Women who work in developed countries are not more likely than women who work in developing countries to have Ph.D.'s.

In many countries, doing one or more postdocs after receiving a Ph.D. is also essential to a successful career. A little more than three out of five of our respondents who have Ph.D.'s worked as postdocs, although postdocs were more common in some countries than in others (Table 8). Of those who did work as postdocs, about half worked in academic settings for their postdocs. About three-fourths of those who worked as postdocs worked four or fewer years.

Among the women who responded to our survey, those from developed countries were much more likely than women from developing countries to have postdocs. Most respondents from developed countries had postdocs, while the majority of respondents from developing countries did not (Table 9).

Academia continues to be a primary employer for physicists. Two-thirds of our respondents are employed in academia. Of these, about two-thirds have tenure or a permanent position. Most respondents from academia say

TABLE 7. Countries Where 40% or More of the Women Physicists Who Responded Do Not Have Ph.D.'s.[a]

China-Beijing

Indonesia

Ireland

Italy

The Netherlands

Nigeria

South Africa

Turkey

[a] List compiled from 29 countries with 10 or more respondents who were not students.

TABLE 8. Countries Where 70% or More of Respondents With Ph.D.'s Had Postdocs.[a]

Belgium

Brazil

Canada

Denmark

Germany

United Kingdom

United States

Countries Where Less Than 45% of Respondents With Ph.D.'s Had Postdocs.[a]

Russia

Yugoslavia

[a] Lists compiled from 15 countries with 10 or more respondents to the second version of the questionnaire.

TABLE 9. Percentages of Responding Women Ph.D. Physicists Who Had Postdocs.

	Country	
Postdoc	Developed	Developing
Yes	73	46
No	27	54
	100%	100%

TABLE 10. How Quickly Have You Progressed in Your Career Compared with Colleagues Who Completed Degrees at the Same Time as You?

	Percent
More quickly	19
About the same	48
More slowly	33
	100%

that it took them the same or less time as their colleagues to get tenure. However, one-third of all respondents felt that they had progressed more slowly in their careers than their colleagues (Table 10). And about one-fifth said that they had less funding and equipment than their colleagues in similar positions or stages.

In addition to earning an advanced degree and possibly taking a postdoc, other steps can be taken in order to have a successful career. Physicists serve on committees, review others' work, and present their research at conferences. The majority of our respondents have given invited talks at conferences. Almost half have acted as referees for journals and have served on important committees at their institutes or companies. Almost two-fifths have served on steering committees for conferences. However, the majority say that they have not served on committees for grant agencies or as editors of journals (Table 11).

TABLE 11. Percentage of Responding Women Physicists Who Have Participated in the Following Activities:

Activity	Percent
Given an invited talk at a conference	52
Acted as referee for a journal	49
Served on important committees	46
Served on a conference steering committee	37
Served on committees for grant agencies	19
Served in an editorial position for a journal	12

Although half of the respondents say that they have served on important committees, it should be noted that women are often excluded from committees that have real power in their institutions and are relegated to less important committees. Regardless of the prestige and power of the committee, women are assigned too often to committees as tokens, and are effectively excluded from having any real input into the decision-making processes of the committee.

"I am convinced that the situation would improve if more women were on the committees dealing with careers, distributions of funding, choosing candidates for a position, etc." [France]

How do respondents think they compare to their colleagues at similar stages in their careers? Almost one-third say that they have served on steering committees for conferences less often than their colleagues have. About one-fourth say they have served as referees for journals and as editors less often than their colleagues. Almost one-third say they have given invited talks less often than their colleagues have.

Marriage

Women physicists who responded to the survey have mixed reactions about marriage. Some see it as a benefit, but many find it detrimental to their careers, especially because it can bring the problem of finding a job near one's spouse. More than one-fourth of the women who responded have never been married, and many chose this route in order to focus on their careers.

"If I would have gotten married, I could have no career." [Spain]

For the women in our study, marriage patterns are different in developed countries than they are in developing countries. Respondents in developing countries are more likely than respondents in developed countries to be married. In fact, one-third of respondents from developed countries are not married, and only one out of five respondents from developing countries are not married. Respondents in developed countries are also more likely to wait until after receiving their final degrees to get married than respondents in developing countries are (Table 12).

Of those who did get married, more than one-third waited until after their final degree to get married, but about half got married during graduate school. About two out of five of those who got married said that marriage affected their work (Table 13). For many the effect was negative, and many of these mentioned the difficulty of finding jobs near their husband's place of employment.

TABLE 12. Timing of Marriage for Women Physicists Who Responded to the Study.

| | Country | |
	Developed %	Developing %
During school	36	59
After final degree	31	21
Never married	33	20
	100%	100%

TABLE 13. Countries Where More Than 50% of Women Physicists Who Responded Reported That Marriage Affected Their Work.[a]

Country	Percent
Egypt	73
United States	60
Russia	59
Germany	57
South Africa	55
Croatia	55
Japan	54
South Korea	53

Countries Where Less Than 30% of Women Physicists Who Responded Reported That Marriage Affected Their Work[a]

Turkey	28
Argentina	26
Indonesia	25
France	24
Ireland	24
Latvia	20
China	19
United Kingdom	19

[a] Lists compiled from 26 countries with 10 or more respondents who have been married.

"I left my tenured position in one of the famous universities of India to join my husband in Brazil in 1975. I waited for two years to get a job in Brazil. The teaching job that I got was in a small university where there were only undergraduate studies." [India]

"I turned down faculty opportunities in Canada to be with my husband in USA. I took a technician job so we could live together. This has slowed down my career." [Canada]

"I delayed leaving the country for a postdoc position and instead started a postdoc in the same group where I did my Ph.D. But finally, since it was not possible to have appropriate positions in the same place in the long run, we got divorced. Only after that I considered to go abroad in order to start a career in science." [Germany]

"I had to leave the US and return to Israel as a postdoc because my husband wanted to come back to Israel. This was after he followed me to the US for my Ph.D. and first postdoc. I still managed to get a postdoc [in the US] in addition to my [other] postdoc position, and to divide my time between Israel and the US. However, I'm certain it would have been better for my career if I stayed continuously in the US." [Israel]

"My work became less important than that of [my] husband. I had to follow his moves, which meant starting again every 3 years." [Belgium]

Other women said that the effect of their marriage was positive because they had married another physicist, or because their spouse was particularly supportive, or because the routine chores of daily life were now shared by two people.

"[Marriage affected me] very positively, my husband has greatly sustained me during difficult periods of my career, both psychologically and by giving moral support, and practically, by taking over more household duties when I had a very heavy workload or pampering me when I had to recover from particularly stressful times. I would never have been able to work as hard and to produce as many papers had he not been constantly on my side." [Belgium]

"I have a lot of support from my husband—he is somebody who would spend an entire weekend in the lab with me if that is what is needed. Without his support, I would definitely not have completed a Ph.D. Of course, a relationship and home life takes time—but I would not change that for anything!" [Argentina]

"[Marriage gives one] something better to do than work through the night." [USA]

"I worked more because he was also in grad school, we encouraged each other." [Brazil]

"I discuss my work at home. My husband is also an astronomer, and a computer wizard, so he helps me by writing programs for me as well." [Netherlands]

"[Marriage] allowed me to devote myself to my studies in a more directed manner: I spent less time at work but improved my working efficiency. Meals and routine chores were shared and became less of a burden." [Canada]

The perception that marriage affected work varied greatly in different countries. In some countries, such as Egypt, the US, and Russia, a majority of married women said that marriage had affected their work. In other countries, such as China-Beijing and the UK, very few women said that marriage had affected their work (Table 13).

Children

In most societies, cultural expectations place most of the responsibility for child care on women. Therefore, most women physicists who have children are typically responsible for child care. They must find ways to balance the demands of their careers with the demands of their family responsibilities. There are cultural pressures to succeed as mothers and wives, and there are scientific pressures to succeed as physicists.

"The main reason, in my opinion, that women physicists rarely reach the highest levels of the professional career is that the society expects them to take (complete) care of the family and children. Our husbands do not like housework at all and they are not ready to share the responsibilities. Hence, spending a lot of time in the house and [on] the children, women physicists have less time to work at the lab. Soon, they become less competitive. Sometimes they have not enough time to follow the new achievements, and slowly they convert to teachers only. Their research suffers." [Bulgaria]

"The odds are stacked against women very subtly and mostly in the form of guilt about neglect of family and children, from family and loved ones at home. That's a combination that's hard to beat. So some kind of awareness among the general public about why it's important for women to hold their jobs and make a mark is also an essential ingredient in getting more women into research, and keeping them there." [India]

"I think many of the problems stem from societal reasons, and these will probably iron out to some extent with time. We have different traditional systems, and they mostly place the women in caring roles (often confined to the home). These coexist with 'modern' roles, and the so-called 'career women' of today spend a good fraction of their energies trying to bridge the divide between what is expected of them at work and at home. Some of the compromises we must make don't earn us much credit in either sphere of action, and can be quite demoralizing!" [India]

"Awareness should be aroused in the society regarding sharing of family responsibilities, so that it becomes easier for women scientists to balance family and career." [Pakistan]

"Since all the burden of raising children falls on the shoulders of the women, there are small chances for them to progress at the same rate as men in the same position." [Israel]

"I think that the situation for all women should be improved as the women are left almost all by themselves to raise the children, work, and take care of the house." [Brazil]

"I believe it is a social problem, not directly related to physics. As long as the housework and children care will not be equally shared by men and women, it will be considerably more difficult for women to be successful in a career, whatever this career is." [France]

Unfortunately, the culture of physics includes the requirement that physicists work long hours. With this comes the related belief that anyone who has to limit her work time is less productive. Women are keenly aware of these beliefs. Some react by concluding that work and family are incompatible and choose to leave physics (unfortunately these women could not be included in this study). Some chose career over family and made the choice not to have children. More than two out of five of our respondents have not had children. Of those older than 45, more than one-fifth have not had children.

Other women have families, but keep putting in the long hours, often returning to the lab after the children are in bed. Still others say that they simply became more productive during the hours they are at work.

"People in science should agree that science is an important part of the life of a scientist—but it's not all of his/her life!!! Everybody in science has the right to have time for family, friends, hobbies. The quality of scientific work of a person does not depend on whether or not this person has also further interests. And scientists should not become slaves of their work because they need or want a new contract or a permanent job at one point. Leaders of research groups, institutes or facilities should not make them slaves. In this sense the culture of work in science has to change a lot." [Italy]

"My husband couldn't understand why I was tired all the time, why I worked so many nights and why I spent most weekends marking tutorial problem sheets." [Australia]

"Having a daughter [while I was] in graduate school gave balance to my life. I spent less time in lab than some of my male colleagues but used my time more efficiently. It definitely changed my work patterns but did not make me less productive." [USA]

"I believe [having a family] taught me to work more effectively, to use my precious time." [Poland]

Among all respondents who had children, the majority waited until after their final degrees to have them. This places most of them into their 30s at the birth of their first child and interrupts research demands just when they are trying to start their careers. The pressure is particularly strong on those who are trying to obtain tenure in academic jobs.

"[Allow] greater flexibility in postdoc to tenure-track, which would allow childbearing without committing career suicide." [USA]

"It would be desirable to have some kind of "sabbatical year" in order to alleviate the pressure for publication and other production for those who have babies." [Argentina]

"I think it would be an advantage if a rule existed that when you apply for a research job, the time you have been on maternity leave doesn't count when they evaluate your list of publications." [Denmark]

Another strain comes from the expectation that physicists make international research visits a priority, as many of our respondents did. Travel is particularly difficult for women with very young children. One respondent reported that a woman she knew had left physics altogether because she was *"asked to go abroad for one year, just after she had a baby."* [Italy]

The cultural expectations to have children are often very different for women in developing countries than they are in developed countries. Respondents from developed countries are much more likely not to have children than those from developing countries.

- More than half of all respondents from developed countries do not have children, but less than one-third of respondents from developing countries do not have children (Table 14).
- Among respondents over age 45, about one-third from developed countries do not have children, but only one in ten from developing countries do not have children.
- Among respondents with children, those from developing countries are much more likely to have them during school than respondents from developed countries are. Respondents from developed countries generally waited until after their final degree to have children (Table 14).

The majority of women with children said that having children affected their work. For example, women with children were more likely to say their career had progressed more slowly compared to their colleagues than women who do not have children (Table 15).

TABLE 14. Timing of First Child for Women Physicists Who Responded to the Study.

	Country	
	Developed (%)	Developing (%)
During school	13	40
After final degree	33	30
No children	54	30
	100%	100%

TABLE 15. Effect of Children on How Quickly Responding Women Physicists Say They Progressed in Their Careers Compared With Their Colleagues.

	Children	
	Yes (%)	No (%)
More quickly	15	23
About the same	45	52
More slowly	40	25
	100%	100%

"Prejudices still exist in the scientific community that a woman who has children usually is not as engaged any more in research like her male colleagues (usually they are fathers and nobody has ever questioned their engagement)...[And there is] the prejudice that the quality of scientific work depends on the time one spends in the lab or office, and the wish of the chiefs to keep their people under control, which means that they do not like (or even believe) that parents work at home sometimes. Many women scientists do not want to live without family/children. Several give up science when they get children as they know, from female friends or colleagues, how hard life is for a mother in physics. Many of those who try to have both, work and family, are soon de-motivated by reality." [Italy]

Evaluation of the Situation of Women in Physics

Although more than three out of four said they would choose physics again, three out of four also said the situation for women in physics in their country needs to be improved. When asked to describe how the situation could be improved, the most frequently mentioned factor had to do with reducing the burden child care places on women. Women mentioned making day care more available or convenient, taking steps to make travel easier during the years when children are young, and having husbands who do their share of child rearing.

The second most frequently mentioned way to improve the situation of women in physics had to do with ending discrimination either across society, or more locally at their individual jobs or schools. These situations are, of course, not unique to physics.

"In the countries I know (Spain and Switzerland), I believe it is not difficult for a woman to become a physicist, if she really wants to. However, it is not true that equality among men and women has been achieved in these countries. I see, for instance, that my daughter (4 years) still receives a considerable pressure at school and from the society to accept the old feminine roles." [Switzerland]

"In my country, as in many others, there is a strong 'machismo.' Men feel diminished if a woman competes with them. Many times they don't mean it, it is just 'the culture.' Some other times, harassment is made on purpose. It is 'always understood' that women are 'frauds,' even compared to the worst male coworker." [Mexico]

"Overall, the difficulty is that women are not seen by men, or in fact by the society as a whole, as likely scientists. This impression influences the attitudes at school and the self-image of young girls and continues to the highest levels when salary is affected for female academics." [Netherlands]

"There are still incidences of sexual harassment, usually in the form of sexual advances that do not stop after verbal requests for them to stop, or illegal questioning about childbearing plans." [USA]

"[The situation for women in physics will improve] if they start treating women as human beings..." [Japan]

"Of course it is also important to make sure that structures are in place to address the problem of discrimination against women. However, from [my] own experience I know that most of the times discrimination takes on a very subtle form. The only way to address it, is to educate women to challenge the culprits. People tend to laugh off laws and official enquiries, however they start to think if they are challenged time and time again by a person." [South Africa]

"In industry, some years ago, I have been subjected to unwanted sexist remarks—(i)size of my [breasts] discussed in front of whole team, (ii) criticized for not wearing make-up as a technical specialist on a trade show booth." [UK]

"I also believe that most of the leading names, certainly in my area of interest, are male and make male decisions. This has enormous consequences for any female young scientist." [Ireland]

"And the attitude that physics is for men must be brushed out!" [Tanzania]

More than two-thirds said that they knew a woman who had left or given up physics. The most frequently mentioned reason for women giving up physics was that they left for family reasons (marriage and/or children). The second most frequently mentioned reason was that they left for a different job, field, or career. Also frequently mentioned was the difficulty of finding a permanent job in physics, which is likely worse for married women.

"[I know women who left physics] simply because they could never get proper positions as physicists, though most of them had the same kind of abilities as their male colleagues." [Japan]

Success in Physics

We asked what factors women thought had contributed to their success in physics. The most frequently mentioned factor was the support of their families, including their parents and husbands. Many also mentioned the support of advisors, professors, and teachers, and some cited the support of colleagues.

"[My success is due to] the support of my husband, who is also a physicist, and who was with [our] kids during my long-lasting experiments or some conferences." [Croatia]

"My family too went all out to support me—my mother with child care and many other things, my husband does at least half the housework and often gives up opportunities if they are inconvenient to my schedule." [India]

"I had the opportunity to work with a very well known professor for a short time. He supported me and encouraged me a lot. It was also psychologically very important for me to see that I could collaborate with one of the best physicists at the time." [Switzerland]

Also frequently mentioned were the women's own determination, will power, and hard work.

"I am very stubborn and when I was told that I could not become an astrophysicist because 'I was not bright enough' or because 'women are not cut out for that' or because 'you cannot have a research career when you have children,' I want to prove them wrong." [Denmark]

"Only my own skills and abilities [contributed to my success]. There was no specific support for women at that time. I think what helped most was my own stubbornness to try things which no one expected me to manage, and see the success after each exam." [Germany]

"I was motivated to work hard because people said the blonde little girl couldn't do physics. I enjoyed the subject and didn't mind working hard at it. I got a lot of satisfaction from working hard and achieving." [USA]

"[There were] stupid (male) lecturers in university who advised me not to work abroad or step over my limits. Therefore, I put even more effort into it, and fortunately I got a lot of support from acquaintances who were convinced of my skills." [Ireland]

CONCLUDING REMARKS

Women pursue careers in physics because they have a passion for the field. They succeed because they are smart, they are determined, and they work hard. In short, they are a remarkably valuable resource for the educational systems and the economies of their countries.

We asked women physicists if they would choose physics again. Three out of four replied that they would. One of the goals of the conference was to identify ways to get this number closer to 100%. Another goal was to identify the changes that will make it possible for the next generation of women physicists to succeed in school and to have the opportunity to successfully pursue careers in physics.

Many women succeed in physics despite tremendous obstacles. All over the world, women physicists report similar barriers to their success. The cultural pressure to be caretakers of children conflicts with the scientific pressures of physics. Women also face barriers in the form of strongly held beliefs that women are incapable of doing good science.

However, women also report similar keys to success regardless of their country. Over and over again, women mentioned the support of others—families, husbands, teachers, advisors, and colleagues—as integral to their success. There is also the continual reference to women's own strength and determination in the face of barriers. These women relied on themselves and on others, and became successful physicists.

"Physics is a subject I love very much and I don't think I can stop learning new things about physics." [Nigeria]

"I think physics is a fantastic thing to do." [Sweden]

ACKNOWLEDGMENTS

The authors gratefully acknowledge the generous support of the U.S. Department of Energy and the National Science Foundation. Their support made this study possible.

The findings in this report are the fruit of a collaborative effort of many individuals and organizations. Special thanks are due to Dr. Marcia C. Barbosa, chair, and the IUPAP Working Group on Women in Physics for their assistance with organizing the study, advice on the questionnaire instrument, and help with data collection. We also acknowledge the team leaders in each country for the essential role they played in disseminating the questionnaires and gathering data on women in physics in their countries. Our deepest gratitude is to the individual women physicists whose generosity with their time and willingness to express their experiences and feelings were essential to this study.

SOURCES FOR COUNTRY-LEVEL DATA AND ANECDOTAL INFORMATION

Albania: Prof. Antoneta Deda.

Armenia: Collected through personal correspondence. Supplied by Dr. Inna G. Aznauryan.

Australia: Compiled from datasets available from the Australian Commonwealth Department of Education, Training and Youth Affairs (Higher Education Division).

Cameroon: Dr. Ndukong Tata Gerard and Dr. Samba Odette Ngano.

Canada: Association of Universities and Colleges of Canada.

Chile: Collected from individual physics departments. Supplied by Dr. Dora Altbir and Dr. M.C. Depassier.

China-Beijing: Dr. Ling-An Wu.

China-Taipei: The Taipei Economic and Cultural Representative Office in Washington, D.C.

Columbia: Instituto Colombiano para el Fomento de la Educacion Superior, and personal correspondence. Supplied by Angela Camacho Beltran.

Croatia: Personal correspondence with physicists at institutions. Supplied by Vjera Lopac.

Czech Republic: Josef Humlicek.

Denmark: *Kandidater i Matematik-, Fysik-og Kemifagene: HVOR GIK DE HEN?* www.uvm.dk/nyt/pm/gik1.pdf.

Estonia: Personal correspondence with physicists at institutions, and homepages of institutions. Supplied by Helle Kaasik.

France: National Ministry of Education.

Germany: Federal Statistical Office Germany.

India: Registrar General of India, Ministry of Human Development.

Iran: Collected from the government and personal correspondence. Supplied by Dr. Azam Iraj-zad.

Israel: Prof. Halina Abramowicz.

Japan: Collected from the Ministry of Education and personal correspondence with another government office. Supplied by Dr. Hidetoshi Fukuyama.

Lithuania: First-level degree information from personal correspondence with universities. Ph.D. information from *Collection of Lithuanian Physicists and Astronomers,* 2nd edition (Lithuanian Physical Society and State Institute of Physics, 2001). Supplied by Alicija Kupliauskiene.

Mexico: First-level data are from direct correspondence with the Asociacion Nacional de Universidades e Instituciones de Educacion Superior (ANUIES). Ph.D. data are from *Anuario Estadistico 1999: Poblacion Escolar de Posgrado,* published by ANUIES.

Norway: Personal correspondence with institutions. Data supplied by Ashild Fredriksen.

The Netherlands: The Dutch Physical Society.

Poland: Central Statistics Office of Poland.

South Africa: Collected through personal correspondence. Supplied by Jaynie Padayache.

South Korea: Embassy of South Korea in Washington, D.C. Supplied by Gayle Juenemann.

Sudan: Collected through personal correspondence. Supplied by Prof. Osman Mai Eltag Mohamed.

Sweden: National Agency for Higher Education.

Switzerland: Swiss Federal Statistical Office.

Turkey: Center for Student Selection and Placement.

Ukraine: *Scientific World.* Supplied by Dr. Oksana Patsahan.

United Kingdom: Higher Education Statistical Agency.

United States: The American Institute of Physics.

Federal Republic of Yugoslavia: Collected by several members of the Yugoslavia working group for physics. Supplied by Dr. Mirjana Popovic-Bozic.

Plenary Papers

Women in Science in Europe:
A Review of National Policies

Teresa Rees

Cardiff University, Wales

Teresa Rees is a professor in the School of Social Sciences, Cardiff University, Wales, specializing in the analysis of education, training, and labor market policies with a gender perspective. She has been Equal Opportunities Commissioner for Wales since 1996 and is an expert adviser to the European Commission on gender equality. She is currently working with the European Commission's Research Directorate on women and science policy in 30 countries (the Helsinki Group). Dr. Rees was the rapporteur for the European Commission's European Technology Assessment Network on women and science that led to the ETAN report in 2000, and is currently rapporteur for the EC's ETAN network on women and science in the private sector. She is a Fellow of the Royal Society of Arts and an academician of the Academy of Learned Societies. Her Ph.D. is in sociology from the University of Wales. She has conducted more than 50 research projects.

INTRODUCTION

There has been growing concern at the European Union (EU) level about the issue of women and science, and more specifically, the underrepresentation of women in scientific careers. This paper describes some of the actions being taken by the Research Directorate of the European Commission (EC) to address the issue, focusing in particular on the work of what has become known as the Helsinki Group. First convened by the Research Directorate in December 1999, the Helsinki Group is made up of representatives from 30 European countries, for the most part senior civil servants who have been seeking to compare national policies, harmonize and benchmark statistics, and develop equality indicators on women in science.

The concern about women and science is not restricted to the EC and senior civil servants. Ministers of Science, senior policy makers, and women scientists themselves have come together at a number of conferences organized by the EC in recent years to debate the issue.[1-3]

The EC adopted a communication in 1999, "Women and Science: Mobilising Women to Enrich European Research," which set out an action plan to promote gender equality in science.[4] The communication committed the EC to mainstreaming gender equality in all its activities. Gender mainstreaming is the systematic integration of gender equality into all policies and programs, and into organizations and their cultures. This led, for example, to the setting of quotas to ensure a better gender balance in EC scientific committees, especially in relation to Framework Programmes. All committees must now have at least 40% of both sexes.

The Women and Science Sector (since upgraded to a Unit) of the Research Directorate commissioned a group of women scientists convened as a European Technology Assessment Network (ETAN) to investigate and report on women and science. This led to the publication, translation, and wide dissemination of what became known as the "ETAN Report" on women and science.[5] The report's title, "Science Policies in the European Union: Promoting Excellence Through Mainstreaming Gender Equality," conveyed the message that women did not want to be treated as special cases—on the contrary. The concern was that the excellence of science in Europe was being compromised by patronage, institutional discrimination, and old-fashioned approaches to human resource management. Examples were provided, including nepotism in the application of peer review systems in the Swedish Medical Research Council and gender pay gaps in the Massachusetts Institute of Technology.

In 1999, the Research Council adopted a resolution inviting member states to engage in dialog and exchange views on national-level policies, taking into account benchmarking and best practices. It also invited member states to establish baseline data on the gender balance of research and development personnel and to explore methods and procedures for the collection of data and the development of indicators to measure the participation of women in research in Europe.

It was as a result of these efforts that the Helsinki Group was set up. But before describing the activities of this group, it is necessary to say something about the "problem" of women in science.

THE PROBLEMS OF WOMEN IN SCIENCE

Despite women enjoying equal access, in legal terms, to education, it is clear that an individual's gender still plays a considerable role in choices made in schools and universities, where the sciences are segregated by gender. Women now make up the overall majority of undergraduates in most western countries. In science subjects they are likely to be the majority studying the biological and medical sciences, but they remain for the most part the minority among physics students and certainly among engineering students. In addition to this horizontal gender segregation, there is clear vertical segregation. Irrespective of national equality laws, scientific infrastructures, or political climates, women remain a very small minority of those at the top of the academic hierarchy, even in subjects where they constitute the majority of undergraduate students. These patterns are extraordinarily robust—across country, across discipline, and over time.

These patterns represent an enormous waste of human resources. Women were found in the ETAN report to "leak" from academic careers at every level of the academic hierarchy. In other words, men are chosen in numbers disproportionate to their representation in the recruitment pool. Women taking a break from academic careers for child rearing find it almost impossible to resume their careers.

The report cites other problems too, including exclusionary mechanisms that keep women out of academic elites, such when incumbents choose successors without using standard, "transparent" recruitment methods. This describes some of our top science jobs and memberships in elite scientific societies, which modern human resource management and equality measures have passed by.

The third dimension to the problem of women in science concerns the gendered nature of the social construction of excellence. Who decides what is worthy of funding or publishing? There is democratic deficit in the decision-making bodies that decide what will be funded and what not, and therefore, what is considered excellent and what not. We should not assume that all decisions regarding science are gender neutral. While universities pride themselves on being based on the principle of merit, they are in fact highly gendered in their organization, from the nomenclature of degrees and the privileging of seniority to the decision-making processes that establish what is regarded as scientific excellence.

THE HELSINKI GROUP

In response to these growing, European-wide concerns about the problems of women in science, in December 1999 the Women and Science Sector of the EC's Research Directorate convened a meeting of civil servants and gender experts from the 15 EU member states and 15 countries associated with the Commission's Fifth Framework Programme. It became known as the Helsinki Group on Women and Science because the first meeting took place in Helsinki during the Finnish presidency of the EU. The group has been meeting in Brussels on a twice-yearly basis ever since. The associated countries are Bulgaria, Cyprus, the Czech Republic, Estonia, Hungary, Iceland, Israel, Latvia, Lithuania, Malta, Norway, Poland, Romania, Slovenia, and Slovakia. All but Iceland, Israel, and Norway are candidate countries for EU membership.

The mandate of the Group is to promote discussion and exchange experiences on measures and policies devised and implemented at the local, regional, national, and European levels to encourage the participation of women in scientific careers and research. It is also charged with producing sex-disaggregated statistics and developing gender-sensitive indicators in order to monitor the participation of European women in research.

The Helsinki Group members have reported on the women-and-science issue in their countries. They have focused on a broad description of the situation and identified specific policies designed to attract women to scientific careers, keep them there, and ensure that fairness mechanisms are in place that will allow them to succeed. This paper provides a synthesis of those national policy reports. The full report was published by the EC in May 2002.[6]

A second group of statistical "correspondents" (delegates) has been set up to harmonize statistics and develop gender indicators among Helsinki Group countries. The EC has also supported Eurogramme to produce national statistical profiles for the 30 countries, summaries of which appear in the synthesis report on national policies.[7] (The Helsinki Group national reports on women and science, the synthesis report, the ETAN report, and the EC conference reports can all be found on the EC Research Directorate's Women and Science Unit website at http://www.cordis.lu/improving/women/reports.htm).

THE CONTEXT

There is considerable diversity among the Helsinki Group countries in terms of scientific infrastructure, equality measures, and the climate for women seeking to pursue scientific careers. Common factors include a lack of gender balance in decision-making about science policy and among those who determine what constitutes "good science." The Helsinki Group has prompted the founding of national committees on women and science in many countries, focusing attention on these issues.

Many countries have instituted affirmative action measures to support women and science. They have supported networks of women scientists, encouraged the development of role-model and mentoring schemes, and in some cases, established targets and quotas. A few countries have experimented with earmarking chairs, research funds, and prizes for girls and women in science.

Gender mainstreaming is supported by the EC in its communication, *Women and Science: Mobilising Women to Enrich European Research.* Nordic countries and some other states have been using gender mainstreaming as an integrated approach to gender equality in all fields. Most countries are using at least some gender mainstreaming tools to help embed gender equality into the systems and structures of science and scientific careers.

Gender mainstreaming tools include legislation. A few Helsinki Group countries have legislation that ensures a gender balance on public bodies such as funding councils. Typically, a minimum of 30% or 40% of both sexes must be represented. Some also insist upon a gender balance on the academic and scientific committees of university and research institutes.

Sex-disaggregated statistics are another gender mainstreaming tool. Candidate countries for EU membership are developing statistical bases that conform to the Frascati Manual and Eurostat conventions. The national statistical profiles of Helsinki Group members are rich sources of data on the position of women in science in all 30 countries. They show how sex-segregation is a feature of scientific careers in all of the countries, although there are variations in the specificity of patterns. The nearer the top of the academic hierarchy, the fewer women there are. Indeed, universally, women are just a tiny minority of women in top scientific jobs.

Gender studies are important for creating a better understanding of the complexities and subtleties of direct—but more particularly indirect and institutional—discrimination. Many countries report support for studies to enhance understanding of the gendering of science and scientific excellence. This has led to a more sophisticated awareness of the use of patronage and nepotism in appointments' procedures, the social construction of "scientific excellence," and the exclusionary mechanisms used by scientific elite bodies.

Other gender-mainstreaming measures that Helsinki Group countries monitor include "engendering" or modernizing human resource management in science, such as by insisting on transparency in recruitment and promotion processes, awareness raising and equality training for those on recruitment panels, and the use of gender mainstreaming experts to advise on gender-proofing policies and practice.

A few countries have identified gender proofing the pedagogy of science education through examples and illustrations used in teaching and textbooks as an approach they are using to identify and eliminate biases in how science is taught. However, while resisting the bog that is biological essentialism, it is important to stress scientific evidence of different gendered modes of learning and setting that against what happens in school classrooms and university lecture theaters.

Measures to facilitate a work/life balance are also crucial to gender mainstreaming. While there is a wide recognition of the issue, progress on addressing it is patchy. Measures reported include good employment practices to facilitate a reasonable work/life balance, and programs targeted at women returners to accommodate their re-entry to scientific careers after a period at home with childcare responsibilities.

Making comparisons across 30 very diverse countries is highly problematic, especially in the case of associated countries that are also candidate countries for membership in the EU. In these countries, in order to fulfil requirements of gender equality, equal treatment legislation is being passed quite quickly to conform to EU directives on equal pay and sex discrimination. In most EU member states, gender-equality policy approaches have developed slowly over the last 30 years, from broad legislation in the 1970s to affirmative action measures in the 1980s to gender mainstreaming in the 1990s and today. However, in candidate countries, examples of all three are being introduced simultaneously. These candidate countries are also experiencing significant changes as a consequence of moving to a market economy, resulting in a rather different role for, and therefore regimes in, universities and research institutes.

FUTURE PRIORITIES AND PERSPECTIVES

Some members of the Helsinki Group acknowledge that by working together and exchanging experiences they have been able to move more quickly on this issue than they might have if they were working in isolation. Future priorities and perspectives, therefore, include facilitating future collaborative working to sustain mutual learning and progress.

A second task entails ensuring more support for research to better understand the gendering of science and scientific careers. Results will be fed into policy development and review.

A third task focuses on the development and use of a series of tools to evaluate and monitor affirmative action and gender mainstreaming measures designed to promote gender equality in science and scientific careers. Scientific rigor needs to be applied to the evaluation of policy approaches designed to promote equality.

Finally, the Helsinki Group is keen to see gender issues mainstreamed into the EC's Sixth Framework Programme. Women scientists from all 30 countries will then have an equal chance of shaping, participating in, monitoring, and evaluating EU-supported scientific projects and programs in the future.

REFERENCES

1. European Commission, *Women and Science,* proceedings of the conference, Brussels, April 28-29, 1998 (Office for Official Publications of the European Communities Luxembourg, 1999).
2. European Commission, *Women and Science: Making Change Happen,* proceedings of the conference, Brussels, April 3-4, 2000 (Office for Official Publications of the European Communities, Luxembourg, 2001).
3. European Commission, *Gender and Research*, proceedings of the conference, Brussels, November 8-9, 2001, (Office for Official Publications of the European Communities, Luxembourg, in press).
4. M. Osborn, T. Rees, M. Bosch, C. Hermann, J. Hilden, A. McLaren, R. Palomba, L. Peltonen, C. Vela, D. Weis, A. Wold, J. Mason, and C. Wenneras, "Science Policies in the European Union—Promoting Excellence Through Mainstreaming Gender Equality," report from the ETAN (European Technology Assessment Network) Expert Working Group on Women and Science (Office for Official Publications of the European Communities, Luxembourg, 2000).
5. Commission of the European Communities, "Women and Science: Mobilising Women to Enrich European Research," COM(99)76 final (Office for Official Publications of the European Commission, Luxembourg, 1999).
6. T. Rees, *The Helsinki Group on Women and Science in Europe: National Policies on Women and Science in Europe in 2002* (Office for Official Publications of the European Communities, Luxembourg, May 2002).
7. T. Rees and A. Parken, "Policies Implemented in Europe to Promote Women in Science: A Synthesis of the National Reports of the Helsinki Group," report to the Helsinki Group (School of Social Sciences, Cardiff University, 2001).

The European Commission Report on Women and Science, and One Frenchwoman's Experience

Claudine Hermann

Physics Department, Ecole Polytechnique

Claudine Hermann is professor of physics at Ecole Polytechnique, the most renowned engineering school in France. Her research domain is optics of solids. She was the first woman to be appointed professor at Ecole Polytechnique, in 1992. Since then, in parallel with her activities in physics, she has studied the circumstances of women scientists in Western Europe and promoted science for girls, by papers and conferences, in France and abroad. She is an alumna of Ecole Normale Supérieure de Jeunes Filles and holds her Ph.D. in solid-state physics.

Based on many unpublished gender-disaggregated statistics, the European Commission's 2000 report on women and science, "Science Policies in the European Union—Promoting Excellence Through Mainstreaming Gender Equality," contains a detailed analysis of the status quo, and ends with practical recommendations for the European Union, its Member States, and their institutions. In this paper I summarize the report, discuss recent French measures addressing women and science, and recount how I became one of the coauthors of the report.

SUMMARY OF THE EUROPEAN COMMISSION "ETAN" REPORT

In 1998 the Research Directorate General of the European Commission set up an expert group with the task of preparing a report on women-and-science policy in the European Union (EU). Growing concern had been expressed at the lack of women, not only among career scientists, but also among those who shape scientific policy. Both the present commissioner for research, Philippe Busquin, and the former commissioner, Edith Cresson, have expressed a strong commitment for gender equality. In the same spirit, the Women and Science sector of the Improving Human Potential Program was created in Autumn 1998. Headed by Nicole Dewandre, it is now a unit in the Science and Society Directorate.

The 2000 report, "Science Policies in the European Union—Promoting Excellence Through Mainstreaming Gender Equality,"[1] was produced within the framework of the European Technology Assessment Network (ETAN) Expert Working Group on Women and Science. It is aimed at "all those whose work has a bearing on educating scientists, creating images of science and scientists, reviewing the work of scientists, recruiting and promoting scientists, funding science, exploiting the results of science and shaping the scientific agenda."

The 14 experts in the ETAN working group were senior scientists from 10 EU countries, working in fields ranging from biology, physics, and geochemistry to social sciences and politics. I was the French participant (and the only physicist). The chair was Mary Osborn, a German biologist, and the rapporteur was Teresa Rees of the United Kingdom (see her paper in these Proceedings). The report represents one year of work and emphasizes collection of gender-disaggregated data and statistics. It is a common point of view between various countries and cultures...and we all learned from regional differences! It concludes that "the underrepresentation of women threatens the goal of science in achieving excellence, as well as being wasteful and unfair."[1] Therefore, the report makes recommendations to a wide range of bodies, including the Commission, the European Parliament, the Member States, and organizations that educate, fund, and employ scientists.

CP628, *Women in Physics: The IUPAP International Conference on Women in Physics,* edited by B. K. Hartline and D. Li
© 2002 American Institute of Physics 0-7354-0074-1/02/$19.00

Status Quo

Table 1 gives the percentage of women in universities in different European countries, ordered by decreasing percentage of women professors; some non-European countries are also listed for comparison. These bulk data include all disciplines, in particular the life sciences, where the percentage of women is much higher than in physics. The ranking raises the question of the prestige of the academic professions in various countries: it is certainly very high in Germany, for example.

In research institutions the percentages of women are somewhat higher than in the universities of the same country: in Italy women represented 13% of the directors of the National Research Centre in 1997, 26% of the senior researchers, and 36% of the researchers. In France, women represented 21% of the research directors at the National Centre for Scientific Research (CNRS) in 1998, 36% of the Chargés de Recherche (lower rank), and 41% of engineers.

TABLE 1. Women Professors: Percentage of Faculty That Are Women (different ranks, all disciplines)

Country	Year	Full Professors	Associate Professors	Assistant Professors
Turkey	96/97	21.5	30.7	28.0
Finland	98	18.4		
Portugal	97	17	36.0	44.0
France	97/98	13.8	34.2	
Spain	95/96	13.2	34.9	30.9
Norway	97	11.7	27.7	37.6
Sweden	97/98	11.0	22.0	45.0
Italy	97	11.0	27.0	40.0
Greece	97/98	9.5	20.3	30.6
Great Britain	96/97	8.5	18.4	33.3
Island	96	8.0	22.0	45.0
Israel	96	7.8	16.0	30.8
Belgium (French speaking)	97	7.0	7.0	18.0
Denmark	97	7.0	19.0	32.0
Ireland	97/98	6.8	7.5	16.3
Austria	99	6.0	7.0	12.0
Germany	98	5.9	11.3	23.8
Switzerland	96	5.7	19.2	25.6
Belgium (Dutch speaking)	98	5.1	10.0	13.1
Netherlands	98	5.0	7.0	20.0
Australia	97	14.0	23.0	40.7
USA	98	13.8	30.0	43.1
Canada	98	12.0		
New Zealand	98	10.4	10.2/23.5	45.5

Gender-disaggregated data are still more difficult to collect for industry. In France, 23% of the 1995 engineering degrees were awarded to women, and the average age of women engineers is 30 years (41 years for men engineers). In the Netherlands women occupy 1.5% of industry's top positions.

In the Academies of Sciences of the Member States, the percentage of women is much lower (1999 data): in France and at the Royal Society of London it is 3.6%. In the Netherlands there is just one woman academician for 237 members; this is to be compared with the U.S. National Academy of Sciences, where 118 of 1904 members are women.

Other Issues

Modernizing Peer Review

The case of C. Wenneras and A. Wold,[2] who demonstrated that a woman had to publish 2.6 times more than a man to be awarded a grant by the Swedish Medical Research Council, is thoroughly analyzed in the ETAN report. Similar biases were evidenced in Denmark. In the Netherlands it appeared that women have a disadvantage in the domains where they are many, an advantage in the fields where they are few. The

Wellcome Trust in the United Kingdom did not find any bias in the attribution to male and female candidates, but noticed that "too few female academic staff are applying for funding."

Educating Scientists, Destereotyping Science

The education of girls has a strong influence on their possible choice of science as a career. Their interest is also determined by their parents, teachers, and career advisers, who all carry stereotypes. The image of science and scientists has to be improved, too. Many European countries advocate role models and mentoring, which should help young people to get a better idea of scientific and technical careers through personal contacts.

"Mainstreaming" Gender Equality In Scientific Institutions and Enterprises

The minimum standard of equal treatment between women and men, as advocated in the EU Amsterdam Treaty, is not always sufficient. In many countries, affirmative actions have been set to compensate for women's handicaps, but they are in some cases considered illegal. The position now highlighted at the European Union level is "mainstreaming," integrating gender in all institutions and actions (see "Women in Science in Europe: A Review of National Policies" by Teresa Rees, elsewhere in these Proceedings).

Gender Statistics In Science

"Reliable, accessible, harmonized data are essential to assess the position of women" in science and technology. The present extreme difficulty with collecting gender-disaggregated data (in particular in industry) is pointed out. Such data, once available, should be utilized in relevant gender indicators (evidencing horizontal, vertical, time segregation). Statistics have to be harmonized, published, and widely disseminated.

Making Change Happen: Recommendations for Specific Measures

An EU directive is needed for gender-disaggregated statistics, as well as for laws in Member States for gender balance in public bodies and access to public records (e.g., to permit analysis of the functioning of the peer review system, as in the Swedish case discussed earlier).

Gender-disaggregated statistical data should then be collected, analyzed, and disseminated.

In the Fifth and Sixth Framework Programmes, the gender dimension in research should be stressed and adequate expertise should be ensured. New specific funding is proposed by the ETAN group for the Sixth Framework Programme to help scientists who cannot spend long periods abroad to build their careers (family responsibilities being obstacles to long absences). Such grants would be awarded to at least 40% of each gender.

The importance of women scientists networking is stressed. A meeting on this topic occurred in July 1999.

EVENTS AFTER THE REPORT

A communication of the Commission (February 1999),[3] resolutions of the Council of Ministers,[4] and working papers[5] dealing with the women-and-science issue have been promulgated. The Fifth Framework Programme contains a target of 40% women on the evaluation panels. In the Sixth Framework Programme, the Science and Society Programme is concerned with gender questions.[6]

The Commission reported on its actions regarding women and science to the European Parliament (December 1999 and Spring 2001). Two conferences took place in Brussels on the progress on this issue (in April 2000 and November 2001).

After the ETAN report was issued, the European Commission's Research Directorate on Women and Science Policy (the Helsinki Group) was created. The missions and activities of the Helsinki Group are described by Teresa Rees in these Proceedings.

In France

Public interest in science education for girls rose in the mid-1990s, partly related to the concern on the poor situation of women in French politics ("parité"). France is ranked among the last of the EU Member States for percentage of women in the national parliament.

Several initiatives in scientific and technical education for girls were proposed in the 1980s and the mid-1990s, in particular detailed statistical reports on the situation of girls on the elite path toward engineering schools.[7] A study of French women physicists was published.[8] The Convention for Equal Opportunities Between Girls and Boys, Women and Men was signed by five ministers.[9] The Directorate of Higher Education at the Ministry of National Education, concerned for several years about the situation of women, is now proposing funding for special actions for women faculty and students in its contracts with the universities.

In 2001 a Mission for Parity in Science and Technology was established at the Ministry of Research; a committee, "Disciplines, Professions, Careers and Gender: The Place of Women at CNRS," was also created.

SHORT BIOGRAPHY OF A FRENCH WOMAN PHYSICIST

Let me briefly introduce myself, to give you a flavor of a French physics career and of how I happened to become a member of the group that produced the European Commission's report on women and science.

I followed the elite French higher education path and graduated from women-only Ecole Normale Supérieure de Jeunes Filles (ENSJF), a school founded in 1881 to educate teachers for the then recently established girls public high schools (the corresponding men Ecole Normale Supérieure de la rue d'Ulm [ENS] was established in 1794). ENSJF became coeducational in 1986 by merging with ENS. Since then, the proportion of women has dropped to 15% in physics and to below 10% in mathematics; it was around one third between 1965 and 1985.

In 1969 I married Jean-Paul, who had followed studies similar to mine at ENS; we have three sons, Philippe, Denis, and Alain; one grandson, Alex, and soon one more. From my mother's example (she was a pharmacist) when I was a child, it was obvious that I would work. In fact, I never stopped working to raise my children, and only took the legal maternity leaves.

I prepared my Ph.D. in solid-state physics in a laboratory at Ecole Polytechnique, the most renowned French engineering school, where I was one of only two women graduate students. (Even then, most other French physics labs had a higher ratio of women graduate students.) Currently the lab has a staff of 50, about half on whom are permanent research staff. I am one of two women on the permanent research staff; however, there are almost as many female as male graduate students working in the lab.

My first academic position (1969–1980) was teaching at ENSJF; all my colleagues were women, so my situation was quite standard. Then I was appointed maître de conférences at the Physics Department of Ecole Polytechnique, and in 1992 I became the first woman professor (the top position) in any discipline since the school was founded in 1794. We are now three women out of about 50 professors, and I remain the only woman professor in physics.

This peculiar situation pushed me to think about the status of women scientists in France. I started to collaborate with Huguette Delavault, a retired university professor in mathematics, who taught me my second research domain, that of women and science. When she was invited to join the EU ETAN group, she proposed that I replace her. I express my gratitude to this great lady, who founded our field in France.

PERSPECTIVES

In my view, the women-and-science issue concerns both men and women scientists[10] and more generally the entire staffs of universities and research institutions. It also concerns the institutions themselves, which should: *(i)* produce gender-disaggregated statistics to be analyzed and disseminated; *(ii)* monitor published texts and illustrations to ensure they give a reasonable place to women; *(iii)* become sensitized to the presence of women, particularly when hiring and promoting; and *(iv)* take specific measures in favor of female staff.

The place of women in science and technology is now a "politically correct" issue. It is thus a favorable time to act and everybody's duty to contribute!

ACKNOWLEDGMENTS

I thank my colleagues from the ETAN group, and Nicole Dewandre of the EU, a permanent source of inspiration on this issue in all the EU Member and Associated States.

REFERENCES

1. European Commission, "Science Policies in the European Union—Promoting Excellence Through Mainstreaming Gender Equality," a report from the ETAN (European Technology Assessment Network) Expert Working Group on Women and Science (Office for Official Publications of the European Communities, Luxembourg, 2000); http://www. cordis.lu/improving/women/documents.htm.

2. C. Wenneras and A. Wold, "Nepotism and sexism in peer review," Nature *347*, 341 (1997).

3. Commission of the European Communities, "Women and Science—Mobilising Women to Enrich European Research," COM(1999)76 final (February 1999); http://www.cordis.lu/improving/women/documents.htm.

4. Council of the European Union, Council Resolution on Science and Society and on Women in Science, 10357/01 RECH87 (June 26, 2001); http://www.cordis.lu/improving/women/documents.htm.

5. European Commission, "Women and Science: The Gender Dimension as a Leverage for Reforming Science," SEC(2001)771 (May 15, 2001); http://www.cordis.lu/improving/ women/documents.htm.

6. European Commission, "Science and Society Action Plan" (Office for Official Publications of the European Communities, Luxembourg, 2002); http://www.cordis.lu/science-society.

7. These reports, under the direction of H. Delavault, are available in French at http://www.edu. polytechnique.fr/Filles/Filles.html.

8. M.-N. Bussac and C. Hermann, Bulletin de la Société Française de Physique (Paris) *114*, 27 (1998)

9. Convention interministérielle pour l'égalité des chances entre les filles et les garçons, les femmes et les hommes dans le système éducatif, Feb. 25, 2000, Bulletin Officiel de l'Education Nationale (Paris) *10* (Sept. 3, 2000); http://www.education.gouv.fr/syst/egalite/conv.htm.

10. See website of the Association Femmes et Sciences: http://www.int-evry.fr/ femmes_et_sciences/.

Women in Physics in Egypt

Karimat El-Sayed

Karimat El-Sayed, professor of solid-state physics at Ain Shams University in Cairo, is holder of the First Order of Merit given by the Egyptian president. She is president of the Egyptian Society of Crystallography and Application, chairperson of the International Union of Crystallography Teaching Commission, and a member of the National Committee of Crystallography and the National Committee of Pure and Applied Sciences. Dr. El-Sayed received her Ph.D. from London University in 1965.

One-third of the total female students enrolled in the higher-education system in Egypt are in science and technology, usually in disciplines such as pharmacy and dentistry. In basic sciences the interest is in the life sciences and chemistry rather than physics or mathematics.[1] Figure 1 shows this trend reflected at Ain Shams University, one of the oldest universities in Egypt. The percentage of women enrolled in some branches of the life sciences is even higher than men.

There is a noticeable change occurring: the numbers of women scientists entering into formerly male-dominated areas such as physics and mathematics are gradually increasing and the gender gap in the annual output of all the faculties of science that exist in Egypt is significantly narrowing.[1] This can be detected from my survey of over 38 women in physics from Ain Shams University of different ages and times of graduation, and from statistics given by the high supreme councils of the universities, shown in Figure 2.[2] Figure 2 shows that the percentages of women compared with men at the levels of professor and assistant professor are much lower than those of women lecturers, who are typically much younger. However, physicists as a whole, both men and women, are not used effectively in the industrial sector because the demand for them in existing industries is rather weak.

Women now make up about 30% of the total number of scientists specializing in physics and mathematics and about 50% of scientists in the life sciences. Women physicists in general, after graduation from universities, have jobs in universities or national institutions, or work as teachers in primary and secondary schools. They do not have jobs in industry. Some married women who encounter great difficulties balancing scientific activities with family duties prefer to leave their careers totally. This is because society forces women more than men to choose between family and career, and for social reasons Egyptian women prefer to have a stable marriage over a promising career.

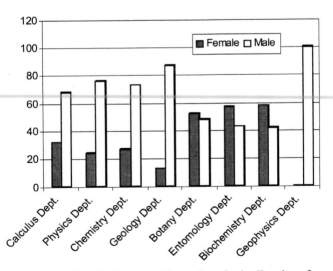

Figure 1. Distribution of staff members in the Faculty of Science at Ain Shams University in 2002.

CP628, *Women in Physics: The IUPAP International Conference on Women in Physics,* edited by B. K. Hartline and D. Li
© 2002 American Institute of Physics 0-7354-0074-1/02/$19.00

81

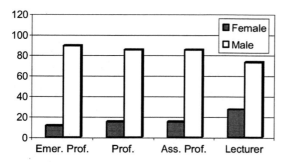

Figure 2. Physics department staff members in all universities in Egypt, 2001.

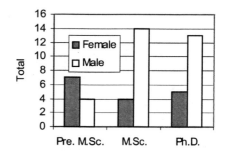

Figure 3. Alexandria University postgraduate students in physics.

From the statistical analysis of women postgraduate physics students at Alexandria University, shown in Figure 3, one can see the decrease in the percentage of women at the M.Sc. and Ph.D. levels, which generally coincide with typical marrying age. But women physicists with appropriate skills have worked their way up, usually aided by selecting an understanding husband who can help them with family duties. These women have reached highly professional, managerial, and decision-making positions. They hold leadership jobs in universities and research institutions, such as dean, vice-dean, department chair, and director. Some are heads of divisions and laboratories; some have been decorated with various orders of merit for their scientific contributions. Egyptian women physicists are also active in nongovernmental organizations.

In general, the percentage of women working in physics is higher in the universities in the north of the country, such as Cairo, Alexandria, and Ain Shams universities, and less in the universities of upper Egypt. This discrepancy is due to the fact that parents in the southern parts of the country do not encourage their daughters to have higher education.

From the analysis of a structured questionnaire developed by the IUPAP Working Group on Women in Physics and given to Egyptian women practicing physics (professors, assistant professors, and lecturers), one can come to the following conclusions:

1. Women physicists are facing a major problem in balancing career and family. A successful woman, after all, is a good wife and a good mother.
2. At both the undergraduate and postgraduate levels, the quality of teaching in physics is similar to other subjects.
3. The choice of physics as a career mainly arises from an interest in physics.
4. Women receive evaluations similar to those of men in physics classes, and woman students are equally recognized by male students.
5. Many distinguished women in physics are taking part in research and education in universities.
6. Many distinguished women in physics are receiving great help from their husbands with family duties and raising children.
7. Most thesis advisors are male, and do not treat women differently than men.
8. Respondents suggested that a certain steps must be taken in order to improve their scientific conditions:
   a. Help with finding a balance between their scientific duties and family responsibilities.
   b. Improvement of their scientific conditions (by developing their research labs, arranging for them to attend conferences and meetings, and supporting them with recent information and knowledge about their work).
   c. Encouraging women, and even men, to specialize in physics by creating jobs that can accommodate them after graduation.

REFERENCES

1. Supreme Council of Universities, Giza, Egypt (personal communication, 2002).
2. Lecture given by the Minister of Education and Higher Education at American University, April 2001.

The Realities of Doing Physics:
Personal Experiences of a Latin American Physicist

Elisa Baggio Saitovitch

Centro Brasileiro de Pesquisas Físicas, Rio de Janeiro, Brazil

Elisa Baggio Saitovitch was a graduate student at the Brazilian Center for Research in Physics (CBPF), where she was awarded a doctorate in experimental solid-state physics in 1973. After her postdoc at Technical University München in Germany, she returned to CBPF, where she has established several experimental facilities for the study of thin films, magnetic intermetallics, high-temperature superconductors, and heavy-fermion materials. The most recent research development in her labs is resistivity measurements under high pressures at temperatures down to 40 mK. She has written more than 200 papers and advised dozens of graduate students. Dr. Baggio Saitovitch has been very active in international scientific collaborations, with special interest in intra-Latin-American groups. She is an associate editor of *Hyperfine Interactions*, vice-president of the International Board of the Mössbauer Effect, a member of the Advisory Board of the Mössbauer Effect Data Center at the University of South Carolina, and vice-president of the Brazilian Physical Society. She will chair the 7th International Conference on Materials and Mechanisms of Superconductivity and High-Temperature Superconductors (M2S-HTSC) in Rio de Janeiro in 2003.

INTRODUCTION

I was asked to talk as a representative of the Latin American physicists. This is for me a great responsibility because within Latin America there is a large diversity in ethnic characteristics, stage of development, and scientific activity. Therefore, I will try to share with you my personal experience as a physicist with the following characteristics: living in a Latin American country, doing experimental research, and being a woman.

The decision to talk about my personal experience as a Latin American physicist was based on a suggestion by our Chairwoman, and usually I like to follow advice, particularly when it comes from a woman. She asked me to not present statistics, since for these you can look at the conference posters about Latin American countries.

Let me make an outline of my presentation. First, I will introduce you to the place where I studied and got my position and to the challenges I have had during my career. Then I will comment about my participation within the Latin-American scientific community. Also, I will talk about how I faced the challenge of reconciling the advances in my career with marriage, motherhood, and family.

I was born in the south of Brazil. I did my undergraduate work in the same university where our Chairwoman is working, Universidade Federal do Rio Grande do Sul. After my degree I moved to Rio de Janeiro, and got my Ph.D. at the Centro Brasileiro de Pesquisas Físicas (CBPF), where I have worked since and am currently a full professor.

Being a scientist in a Latin American country, one has to face ethical dilemmas that some other countries do not have. Science is financed by the government, which has other urgent problems to address. In a country with serious problems such as hunger, lack of housing, and a high percentage of infant death, scientific research seems to be a luxury. The funds designated to science are small and are the first to be cut when there are economic problems.

However, we know, and we have to help the government and the overall population realize, that we only will be able to overcome our country's problems with development—and that to achieve development, science is essential. The country's structural problems cannot be eliminated without real development. From the perspective of globalization (which indeed is nothing but a new name for the imperialism of rich countries), the only way to achieve development is by educating the overall population. This will enable the majority of the people to handle technological innovations, and create an elite with the ability to understand these technologies, to adapt them to our needs, and to create new ones.

CP628, *Women in Physics: The IUPAP International Conference on Women in Physics,* edited by B. K. Hartline and D. Li
© 2002 American Institute of Physics 0-7354-0074-1/02/$19.00

Sometimes when searching for funds to do experiments, we get stuck because of economic problems and our government's failure to understand the need for scientific research. We wonder how much farther we should go, and we may consider giving up. Then we remember just how important for the development of our country is the innovation we are generating and the education we are giving to the people and decide to keep struggling.

To illustrate this point, let me tell you a story that actually happened with my advisor. Once he was in Guatemala visiting some historical ruins. His guide began to take stones out of the ground. He reprimanded the guide, telling him not to take the stones, that they were ancient and historical. The guide replied: "Señor, como quiere usted que yo respecte las piedras? Tengo hambre y en el pueblo mi familia no tiene ni mismo lo que comer. Como quiere usted que se respecten las piedras, si acá no se respectan ni mismo los hombres?" (Sir, how I can respect the stones if I and my family are hungry and have nothing to eat. How do you want anyone here to respect these stones, if in this country, even man is not respected.)

CBPF was founded 50 years ago as a nongovernmental organization by some young and brilliant Brazilian physicists returning from abroad (Cesar Lattes, Leite Lopes, and Jaime Tyomno). It now belongs to the Ministry of Science and Technology. It was created with support from intellectuals, politicians, and industrials to be a center for scientific achievements (at that time it was not possible to do research at the universities in Brazil). CBPF has significant stature in the Brazilian and Latin American scientific community, because many in this community received their degrees from CBPF. Indeed, the institute has a long tradition of giving fellowships to students from other Latin American countries. We now have about 30% female researchers with permanent positions.

THE CHALLENGES

What are the challenges of doing research in Brazil? Many of the challenges women physicists face are also faced by men physicists, but others are not. We are located far from the most important centers in science and physics, and this constraint needs to be overcome by collaborative programs and participation in international conferences. Many trips can be very stressful for a woman who leaves the family behind, but it is necessary. I have colleagues who could not pursue postdoctoral work because their husbands, for all practical purposes, would not allow it. Needless to say, this is very bad for their careers because it hinders their ability to be promoted.

There are many other difficulties related to getting a job, because the only career possibilities for physicists are to teach or do research. There is almost no opportunity to work in a company or in industry, because the important ones are multinationals that prefer to have their laboratories in Europe or the United States.

We have a problem with funds, mainly funds for experimental research. But we must compete, we must fight. We must submit proposals. It is very stressful task, but we cannot give up. This is true regardless of gender.

It may help if we pursue less-explored research areas, because then we can compete more easily. On the other hand, this is less challenging and leaves the scientific community out of the mainstream. I was advised to do just that 10 years ago, but did not, and am glad for making that decision. Our institute was being evaluated by a commission of eight physicists, and one of them suggested we should not work on high-T_c superconductors because that was a very competitive area. The commission, however, was supporting our other proposal, which was for research on multilayer materials. Even knowing that the money would not come easily, I continued to work in high-T_c as well, and since then we have published many papers and completed several Ph.D. thesis on high-T_c.

Another result of my persistence is that in May 2003 I will organize the next International Conference on Materials and Mechanisms of High-Temperature Superconductors (M2S-HTSC-VII) in Rio de Janeiro.

To work in experimental physics you need to be able to build a group. Over the years I was able to build a team that now has more than 20 people—20% of them women—including students, postdoctoral fellows, and senior researchers. The only team member besides me to have a permanent position is a woman. She received this position just two months after having her baby. Balancing her family responsibilities with her research was particularly challenging during the baby's first year; of course, this generated some problems for her during that time. But this period has passed and the balancing is going smoothly now. When I hire someone for the group, I do not pay attention to whether the candidate is a woman or a man, but somehow I end up having many women working with me. Like the other members of my group, they benefit from my large international collaboration with physicists in Europe, the U.S., Japan, and Latin America.

To keep this structure and many projects running, I have to search for funds. I also need to be persistent and to pay attention to policies and procedures, otherwise I may end up without proper financial support. So, I have been in this trajectory for many years, and have received support for my work. Particularly, one of my programs, the Highly Correlated Electron System, is funded by PRONEX, a program designated to finance "Centers of Excellence" in

Brazil. Aside from running research, in Latin-America scientists also have to be involved in keeping and improving the infrastructure facilities. I have been for 20 years in charge of the low-temperature facilities of our institute.

It is very important to be an active member of the Latin-American scientific community at large. From 1984 to 1990, I was a member of the evaluation board for the Latin American Center for Physics, an intergovernmental institution created in 1952, whose main duty was to determine whether this kind of institution still had a role to play in Latin America. Through this experience I realized that the scientists in our region have very little interaction and collaboration. The Argentineans tend to collaborate with the United States, and the Brazilians lean toward Europe and the U.S. We need to promote regional collaboration, which has the potential to enhance our research capability.

Some other scientists and I have been working in this direction, trying to improve intra-Latin-American collaboration. In 1988, I began a series of Latin American Conferences on Mössbauer spectroscopy; the next meeting of the series will be held in Panama this year. Additionally, I have many students from throughout Latin America. Some of these students, after working for me, go back for their countries and have to face poor working conditions. I have collaborations with Cuba, Peru, and other countries.

As leader of the M2S-HTSC conference, I wanted it to have a Latin-American organization, and I invited an Argentinean to be the chairman of the program committee. Recently, as you know, the economic situation in Argentina became very bad forcing him to decline the invitation. I was so moved by the situation of the Argentinean scientific community that I called the secretary of the Minister of Science Technology. The conversation was very much a contact between two women. I suggested that Brazil should establish a program to support science in Argentina similar to the one by Europe to help the former Soviet Union. There were many reasons for proposing the program, among them the fact that I was already much involved with Latin-American issues in the area of physics. The program would allow students and researchers to work in Brazil for a short period without losing their affiliation with their home institution in Argentina. After their return, they would also receive support for pursuing this collaboration. She transmitted my idea to the Minister, who became enthusiastic and asked me to write a detailed proposal for the project. So I worked until 1 a.m. and sent by e-mail a one-page proposal. This proposal is the basis for the large exchange program that now exists between Brazil and Argentina.

Our institute has a good graduate school. Looking over the more than 500 thesis we have produced over these 52 years, I notice that we have more women pursuing their doctorate degrees than their master's. I believe this trend reflects the fact that it is easier for males to travel abroad for their doctorate degree. I would also like to stress that our institute has a long tradition of giving fellowships to foreign students. Interested parties should contact us.

MY EXPERIENCE AS A WOMAN IN PHYSICS

Some of my own experiences illustrate the prejudice the international community exhibits toward women in Latin America. At the beginning of my high-T_C work, I attended a conference to present my results. I had, however, results completely different from those presented by the other physicists. An invited speaker from the U.S. came to my poster and looked at my results, probably thinking: "Can I believe these results from a Brazilian woman?" But he did not say anything. On the last day of the conference he came to me with a copy of his paper and said, "I think you were correct. I am going back to my lab and checking my samples."

In another conference that took place in Brazil, after listening to a discussion I had with one scientist, an American physicist came to me and said: "I would not like to get a divorce from you. It is difficult to debate with you." On another occasion, another physicist came to me and said: "I would not like to be your husband. It would be very hard to live with you since you travel too much and you are too aggressive in discussion."

On the other hand, I have also had some nice experiences. In 1989 I gave a talk in Tokyo at a University for women and it was a very nice feeling just to be there together with them.

But now, let me come to the main point of this conference. I cannot tell you that I felt discriminated against for being a woman when I was in school and at the university. Maybe I was not sensitive enough to it, or maybe I did not experience it because I was a very good student, and if a woman is a good student it is more difficult to discriminate against her. However, at the career level, I have perceived some signs of discrimination. For instance, I was the first woman in my institute to become a full professor, 30 years after its foundation. When the institute was created it had two female researchers, both of whom were very active, but they never attained the position of full professor. Another touch of discrimination can be seen in the research grants. We have in Brazil a fellowship that is granted according to one's research level. These levels are defined according to productivity and ability to advise students, lead research, etc. Although my productivity (numbers of students, publications, and so on) easily matches or exceeds the average, I was never promoted to the first research level by the selection committee, which so far has been composed only of males. The Brazilian Physical Society once profiled the average recipient in each of the research levels. In all of the measures my profile fits the first level.

I have very often been on a committee where I am the only woman. Some women might not realize this, but as the only woman on a committee, you might be confronted with some difficulties. For example, when I proposed to organize the next conference on High-Tc in Brazil, there was a committee made of 50 men and me. Even feeling unease as the only female scientist in this committee, I went on with the project. We are so few and so discriminated against, that when we get an opportunity we have to accept the challenge.

MY LIFE IN PHYSICS

Now I would like to tell you something about how I came to work in physics. In Brazil, when I was young, there were two different types of high schools: one for students' inclined to go into the humanities and the other for students in science. I wanted to become a lawyer, so I went to the first type. However, after the second year I decided that I wanted to study science so I moved to the other type. When I finished my high school studies and I had to decide whether to concentrate on physics or math, I opted for physics. When I finished my undergraduate studies, I was confronted with another choice: experimental or a theoretical? I went for experimental physics.

After my degree I married a colleague physicist, and I went to Rio de Janeiro to pursue my graduate studies at CBPF. I was very fortunate because I was able get a permanent position there without yet having my doctorate degree. This is impossible today.

My advisor Jacques Danon was a very special man. He came to Paris in 1949 to study Philosophy after his degree in chemistry. Soon being in contact with intellectual groups he met Joliot and Irene Curie and they convinced him to study radiochemistry. He married a French biologist and spent his life between Rio and Paris. He was a man who already had many women working with him 20 to 30 years ago. He was a humanist, very sensitive, very creative, and he worked on interdisciplinary topics. Unfortunately, he is no longer living. I think he was one of the most important persons in my career.

When I was doing my postdoctoral work in Germany I decided to have a child. But life is a bit more complicated. In seven years I became pregnant five times, and I lost the babies. This period was a very difficult time for me. After trying so many times, my husband wanted to give up. He said that if I had already lost five, I would probably lose the sixth. I decided that we could not give up and finally we had two girls, Anna and Flora. During this period, I also became more connected with women's issues and active in the women's movement. In Rio de Janeiro we created a movement, "SOS Woman," aimed at helping abused women. This was very important for me, because before that I was the kind of woman that felt more comfortable being among men than among women. For instance, if I was at a party, I always would talk with the men, because they had, in my understanding at that time, a more interesting conversation. However, through my participation in the women's movement and my painful experience of pregnancy, somehow I recovered the connection with my female side as a mother. The fact that I wanted to become a mother and I had a problem achieving that goal changed my perspective toward womanhood.

Being a mother and a scientist is not an easy task. When my older daughter was an infant, my husband and I brought her to a Brazilian national meeting on condensed matter physics. And what happened? I stayed with the baby while my husband went to the talks. So for the next conference, and maybe for the next 10 years, I went with Anna and my second baby, Flora, and tried to get some help during the week with caring for the children. After the meeting was over, my family and I would stay on in this same city to have a weekend together.

Sometimes in our careers we are overloaded with work, and reconciling it with taking care of the children is difficult. We have to help our children understand the value we give to our work. When I started my work in high-Tc, a very competitive field, I was working very hard. Sometimes I had to bring the children home at 5 p.m. from their school, feed them, put them to sleep, and go back to the lab and work until 2 a.m. The next morning at 6, I would get us all up and start over again. One day, while my husband was traveling, I felt that I had to explain to my kids why I had to work so hard and be gone from home for so many hours. I said to them: "You like ice cream a lot, right?" They said, "Of course we like ice cream!" I said, "For me, my work is like ice cream is for you. The more I work, the more I want to do." Each of my daughters reacted differently, however both understood the idea. The younger one suffered very much with my travels, but now that she is 16 years old, she accepts my need to work. The older one always accepted it with no complaint and actually she wants to be a physicist (she is in her first year at college). Of course, the younger complains, since she is now surrounded by three physicists.

Being a working mother gives you a feeling of guilt. You don't know whether to go to your children or to your work. I think every woman who is really involved with a career feels this conflict. But we should try to handle it.

Finally, I would like to talk about by family background. My grandparents emigrated from Italy to Brazil and lived in the countryside, in the south. I was the seventh in a family of eight children (five boys and three girls). My father never made any distinction between boys and girls regarding education. He always said, "The only thing I can give to you that nobody can take away is your studies." As a result of this attitude, both my older sister and I became scientists. She works in genetics, and she is very active. In fact, she also has one project approved for the Centers of Excellence program. We once discussed why we both are so work-oriented, and we believe that it actually is not our father who has influenced us in this respect, but our mother. Our mother worked very hard. While she raised all these children, she was also working in the family shop day and night. Somehow I think this has passed to us.

I would like to add that motherhood always perturbs a career. For example, if you have a child after earning your Ph.D., it is difficult. If you have a child while you are preparing your thesis, it is difficult. If you have a child when you are older or closer to being a professor, it is also difficult. In all cases you demand a lot of yourself and have many goals you want to reach. In a developing country, of course, it is common and easy to have the help of a maid at home. Even so, we have other difficulties with balancing family and career. In retrospect, I wouldn't consider forgoing the possibility of having a child, even if you really love your career. Indeed, children and family have been something very, very important to me.

We women need to contribute professionally—to the advancement of physics and knowledge—and also to our society by being mothers to children. In both arenas we can achieve our aspirations. My daughter now is studying physics, and this is a good contribution in terms of the topic of this conference.

Status of Women Physicists in Japan—Present and Future

Masako Bando

Aichi University

Masako Bando is professor of physics at Aichi University. She specializes in high-energy physics, particularly grand unified theories and neutrino physics. She is the former dean of the Faculty of Liberal Arts and Science at Aichi University. Dr. Bando was instrumental in creating the Kyoto University's daycare program when she was a Ph.D. student there in 1960s. She has authored or coauthored three books and more than 150 papers.

HISTORY OF WOMEN SCIENTISTS

Japanese universities that had previously accepted only male students enrolled female students for the first time after World War II. Since then the number of female students has gradually increased and the status of women scientists has greatly improved. Yet, statistics show that the number of women researchers today is still very modest. Women faculty members in all fields at universities are 10%, and even lower for professors.

The history of women scientists in Japan can be divided into three stages:

1. *1956–1975: Opening of Child-Care Centers*
   The number of women workers and, accordingly, the number of women researchers rapidly increased in this period. Encouraged by the demands of women workers for more child-care facilities, women researchers organized movements to request child-care facilities at universities.

2. *1975–1985: Declaration of the International Women's Decade*
   The International Women's Decade encouraged the women researchers' movement and created a climate for exploring issues concerning women researchers on a national scale. The Japan Science Association (JSA) organized an annual symposium on this issue. The Science Council of Japan (SCJ) created a subcommittee to improve the status of women researchers and submitted appeals to the government concerning women researchers' status. The SCJ subcommittee researched the life-cycles of women scientists. They performed statistical analysis of questionnaires and compared research performance and life-cycles of women with those of men.

3. *1985–2001: Effects of Equal Employment Opportunity Law*
   Japan finally passed the Equal Employment Opportunity Law near the end of the International Women's Decade. Encouraged by this law, the female members of SCJ organized JAICOWS (Japan Association for Improvement of Conditions of Women Scientists) and started to investigate the status of women scientists. In 2000 the Japan Association of National Universities (JANU) issued a report on the status of women scientists and proposed that the national universities should endeavor to increase the number of women academics at national universities.

FINDINGS OF JPS AND JSAP INVESTIGATIONS

In preparation for the 2002 IUPAP International Conference on Women in Physics (these Proceedings), the Japan Society of Applied Physics (JSAP) and the Physical Society of Japan (JPS) organized committees aimed at the equal partnership of men and women in academic society, and launched a joint project that included investigations by questionnaire (see "Summary of Nationwide Survey: The Work Environment of Physicists in

CP628, *Women in Physics: The IUPAP International Conference on Women in Physics,* edited by B. K. Hartline and D. Li
© 2002 American Institute of Physics 0-7354-0074-1/02/$19.00

Japan," elsewhere in these Proceedings). Thirteen percent (2619) of JPS members and 16% (3604) of JSAP members responded to the questionnaire.[1,2] I comment here on the results.

By dividing responses into academia and industry, we can clearly identify what affects the status of women physicists. Studying the effects separately was possible because the majority of JPS members (60%) belong to academia, while members of JSAP (56%) mostly belong to industry. We find that in industry the status of women was dramatically improved by political changes such as the Equal Employment Opportunity Law or the Child-Care Leave System. This proves that once women are in better working conditions, they contribute equally to scientific development while balancing family life.

On the other hand, in academia we find no appreciable effect of these political changes. The status of women physicists was better only in periods when the absolute number of academic staff increased and only in the specific areas where child-care centers were available. The JPS data survey found women physicists who managed to continue both work and family-care even into their 60s. Most of them (including myself) benefited from a relatively good employment market reflecting the "science boom," and from spending their motherhood years in an area with good child-care facilities.

When I had my child, I was in the Ph.D. program at Kyoto University. I did not want to give up research in order to be a good mother, and made up my mind to improve conditions. Some other women scientists and I organized a movement to construct a daycare center at the university. During this period, my house was a daycare center, which the other women and I managed by ourselves for two years. Now in Kyoto University there are two good daycare centers with excellent staffs. We are very proud of this.

Currently the situation is different in academia than in industry. There is less chance for a woman to gain a position, especially in her early years as a researcher, because of lower research performance due to motherhood responsibilities. The Child-Care Leave System does not work well in its present form without good child-care centers and support by peers in the workplace.

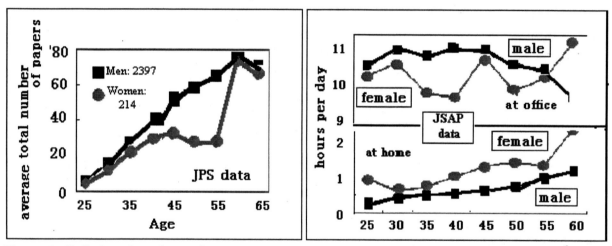

Figure 1. (*Left*) Age dependence of average total number of papers for male and female (JPS data).[2] (*Right*) Average working hours at home and in industry of males and females (JSAP data; figure reprinted with permission).[1]

JPS data show a representation of research performance gauged by the numbers of accumulated published papers (Figure 1). We observe that the age distribution of the averaged number of publications of women deviates from men's at around 35 years of age. Note that the average number again approaches men's at around 60 years of age. The activity of women seems to quickly recover after the "child-care period." Let us take the hypothesis that there exists a period in the life-cycles of women physicists in which they become brilliant after domestic obligation, the "recovery phase." We shall return to this point later.

Other data from the JSAP study (Figure 1) are suggestive, showing how many hours per day women and men work at office and perform career-related work at home. No meaningful statistical difference was found in the amount, until about age 60. Also note that a more careful look at Figure1 leads us to conclude that women's working hours both at office and at home far exceed those of men around the age of 60. This again may be a confirmation of the hypothesis of a recovery phase in women's life-cycles.

LIFE-CYCLE FROM MICROSCOPIC AND MACROSCOPIC VIEWS

In order to identify the barriers for women physicists more clearly, I discuss life-cycle analysis in terms of the index of research activity, which is gauged by the annual number of publications. From Figure 1 we can estimate this as $N(y + 5) - N(y)$, where $N(y)$ is averaged accumulated published papers of the people between y and $y + 5$ years old. It is approximately the publication numbers during every 5 years with which we draw the typical male and female life-cycles of research activity (Figure 2).

This macroscopic profile is interesting. We recognize three phases in the life-cycle of women physicists; the early stage is almost the same as men, and after the second stage from the mid-30s to the mid-50s, they enter into the recovery phase, in which activity increases greatly. Of course, this picture may not be universal because the number of respondents aged mid-30s to mid-50s is very low. Nevertheless, it may be instructive and worth further study, if we look at various data very carefully.

I wish to propose a microscopic approach. Cole and Zuckerman interviewed 120 scientists in USA to see whether marriage and parenthood are incompatible with a scientific career. Among them was an eminent woman scientist with three children, whose publication history is illustrated in Figure 2.[3] The profile looks quite similar to the macroscopic plot for women shown in Figure 2. Of course, the research performance of a physicist depends on

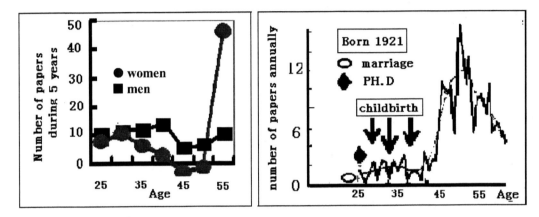

Figure 2. Profiles of activity from (*Left*) macroscopic view comparing men and women, and (*Right*) microscopic example of a woman researcher with three children interviewed by Cole and Zuckerman (in USA; figure reprinted with permission).[3]

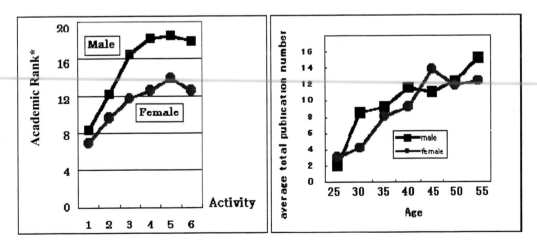

Figure 3. (*Left*) Academic rank* vs. activity index.[4] (*Right*) Average number of publications as a function of age.[4] (*Academic rank is a combined measure of institutional status [e.g., national university, private university, college] and professional rank [e.g., professor, assistant professor, research associate]. The term is explained in more detail in ref. 4.)

each life history and reflects various surrounding conditions. It would certainly be possible to extract meaningful elements if we collect life-cycle data of men and women physicists in various countries. They will give us definite information on the hypothesis of the recovery phase. I would like to ask you to draw your life history like that demonstrated by Figure 2, with some comments (marriage, childbirth, Ph.D., etc.). By combining macroscopic data with microscopic data, we can extract barriers for women to continue research and explore ways to make excellent achievements in physics with a good balance of family life.

There were previous projects to investigate the status of women researchers in Japan. JSC committee members did the first in 1982,[4] followed by JAICOWS in 1998.[5] The results showed that men enjoy much higher rank than women with the same activity index (Figure 3). This clearly refuted a popular explanation that women are in lower positions because of their being less academically active than men. These results also articulated that the most serious difficulty is the child-care issue. Indeed, by dividing answers of natural scientists into those with and without children, it is confirmed that lower research performance is more prominent for women with children. We see similar profiles from the age dependence of the total number of publications of natural scientists (Figure 3). Note that Figure 3 indicates the recovery phase again. This is confirmed by the JAICOWS data: the average activity index of women decreases according to the increase in the number of children, while men show the opposite tendency.

These results may seem to contradict to the conclusion of Cole and Zuckerman.[3] They claim that marriage and family obligations do not necessarily affect the scientific productivity of a woman. However, if we look at their argument, we can extract the necessary conditions for women scientists to overcome the dual roles. Because thinking about science can occur at home as well as the workplace, and if she has a scientist husband, a woman scientist can have opportunities to talk about her research during so-called off-time. If she has good collaborators and few professional obligations other than research, she can stay active even during the child-care period. Of course, we must be very careful in making a final conclusion: a lower rate of publication in the early years is not necessarily attributed to the demands of motherhood, but may instead be characteristic of the beginning phase of a developing research program.

The above argument is instructive because it suggests the importance of socialization into a profession."[6] Women tend to be less motivated to develop their academic careers because of a lack of opportunities for "socialization into a profession" caused by the social gender bias, and are often outside of closed and hierarchical structures. Making opportunities for socialization into a profession as equal as possible between men and women is called for. Looking at "socialization of women into a profession" critically and exploring how to change the current closed, unfair system into a more open, merit-based system, may be crucial for increasing the ratio of women researchers.

I believe that women as well as men want to improve the present recruiting system, especially to realize the evaluation system for employment based on the principle of openness, transparency, and fairness. Both men and women share the benefit if we can improve the conditions of academic society, the employment system, and the decision-making system.

Finally, I wish to comment on possible important roles women researchers might take. Women in many cases experience more multi-style lives than men. Many of them spend less time in pure scientific work than men, but they are more concerned with family care and daily life than men. Such experiences of women may contribute to the creation of new areas in science. I believe that the quality of science and the atmosphere of academic society will improve when men and women appreciate different experiences and perspectives, share them, and work cooperatively. Science shared by men and women equally would enhance its diversity. This may be the motive force to develop science and technology, yielding a reconstruction of human culture.

ACKNOWLEDGMENTS

This report is based on the joint project of JPS and JSAP. I am grateful to the organizers of JPS and JSAP for their effort to perform the project. I wish to thank Prof. A. Kosaka of Aichi University, Prof. A. Ito, and Dr. M. Hatsuda for their valuable comments. Thanks are also due to Y. Enyo, T. Kagayama, M. Fujita, and M. Toya for their collaboration in writing this report and M. O. Watanabe and E. Tamechika for providing the figures for JSAP.

REFERENCES

1. M.O. Watanabe, E. Tamechika, K. Domen, and Y. Okada, "Survey of investigation on present status of JSAP members" (in Japanese), Oyo Buturi *71* (5) (2002) 510.

2. A. Ito, Y. Enyo, Y. Ogushi, T. Kagayam, M. Toya, E. Torigai, K. Nomura, M. Bando, and Y. Fujita, "Report of analysis conducted by questionnaire" (in Japanese), Butsuri *57* (2002) in press.

3. J.R. Cole and H. Zuckerman, "Marriage, motherhood and research performance in science," Scientific American *17* (4) (1987).

4. K. Shiota and K. Saruhashi (editors), "Life cycle analysis of women researchers in Japan" (in Japanese), Domes (1985). See its abstract in English in: M. Bando, Y. Kozai, and M. Toya, "An investigation of the status of Japanese women scientists," AWIS Newsletter *XVIII* (3) (May/June 1988).

5. H. Hara (editor), "Status of careers of women scientists in Japan" (in Japanese), Keiso (1998).

6. M.S. White, "Psychological and social barriers to women in science," Science *170*, 413 (1970). The PowerPoint file of this talk may be seen on the Web site: http://www.leo.aichi-u.ac.jp/~bando/contents.html, where more detailed contents including various figures are also available.

Russian Women in Physics

Iya Pavlovna Ipatova

Ioffe Institute, Russian Academy of Sciences

Iya Pavlovna Ipatova is a principal researcher at the A.F. Ioffe Physico-Technical Institute of the Russian Academy of Sciences and professor of physics at Technical University in St. Petersburg. She received her Ph.D. in physics from St. Petersburg State University. Her areas of research include lattice dynamics of pure and defect crystals, disordered systems, physics of the surface, light scattering from semiconductors, and theory of physical processes in technology of semiconductors. Dr. Ipatova has authored or coauthored more than 300 publications, including four textbooks. She is a member of five national councils of the Russian Academy of Sciences.

PERSONAL ASPECTS

I am happy to have the good fortune to be coming to the Ioffe Institute every day for more than 40 years. The beginning of my career was very fortunate. I had an excellent teacher. Because he was blind, we had a unique teacher-student experience. Whatever research we did, we did together. I truly appreciated his erudition, his ability to pick out a problem of interest and find a mathematical solution, analyze the results, write papers, and eventually report the results at conference. I always felt his support behind me until he realized that I could work on my own.

He belonged to a generation of physicists that I would call romantics—he loved physics selflessly. He taught dozens of students in St. Petersburg, among them two women who would later work on their own. His name was Lev Gurevich, and he belonged to Landau school of theoretical physics in Russia. There are two effects which make his name immortal: drug effects in solids and negative pressure in stars.

Because of my connection with Lev Gurevich, whatever I did at the beginning of my research career was recognized—I was protected by his name and authority. But usually the scientific results of women physicists meet a very low level of recognition. It is unfortunately common for a woman scientist to find her results underrated and belittled. It is one important reason that to be a physicist is not prestigious for a woman today, with the result that many talented and ambitious women give up their research careers in physics.

NONRECOGNITION OF FEMININE ACHIEVEMENTS IN PHYSICS

The story of Lise Meitner, German physicist who contributed much to the development of nuclear physics, is well known. Books have been written about Lise. The magazine "Physics Today" devoted issues to her life and achievements. But her lesson has not been learned yet.

I am sure that you have never heard the name of Nina Goryunova from Ioffe Institute. She is the creator of semiconductor alloys, mixed crystals of the types $A_{(1-x)} B_x C$ and $A_{(1-x)} B_x C_{(1-y)} D_y$. The materials are the basis of heterostructures that were awarded the Nobel Prize in 2000. But her name is very seldom mentioned, if at all, by the men speakers, even in our institute, where she worked.

Another example is my colleague Olga Sreseli, who made beautiful measurements of surface polariton spectra in semiconductors more then 10 years ago. But when she tried to say to her boss that this work could be the basis of her thesis, his reaction was negative. Olga finally received her Doctor of Sciences degree, but there were an unhappy 10 years in between.

Quite recently I asked the head of the laboratory in our institute how the two young postgraduate women in his laboratory were. His reply: "Quite well. One of them is a bookkeeper in the lab and the other is doing measurements on technological setup." I asked whether she ever came up with any interesting ideas and whether she has the chance

CP628, *Women in Physics: The IUPAP International Conference on Women in Physics,* edited by B. K. Hartline and D. Li
© 2002 American Institute of Physics 0-7354-0074-1/02/$19.00

to learn the physics of the experiment. "No, it is not required of her," was his reply. These words could ruin the career of the young physicist.

What can we do with this lack of trust and attention?

I have found an outstanding example given by American women physicists from Los Alamos. They are Ruth H. Howers and Caroline L. Herzenberg, who published the book, *Their Day in the Sun: Women of the Manhattan Project*. In memoirs about the Manhattan Project published around the world, there are women as wives and mothers but not as physicists. We now know from this book that there were women physicists on the project. The book alone should get honest readers to think that the community of physicists didn't recognize the women of the Manhattan Project until now.

IMPROVING THE CLIMATE FOR WOMEN IN PHYSICS

Creation of a comfortable and respectful atmosphere for the creative work of gifted women is a very important aspect of improving the experience of women in physics. Women are different than men psychologically. Many of us lack self-confidence. The reason lies in the nature of girls' upbringing in the family. I had never met a man in charge of a project who would be worried that the women on his team may lack self-confidence. They say: "Why create problems? I am too busy to bother." This same behavior is expected from women scientists. It is often the source of conflict. This is the reason why it is very important to involve men in our discussions about improving the climate for women in physics.

Nevertheless, an experienced department head is supposed to try to obtain the most creativity from each member of the team. Women are still viewed as the intellectual reserve of the 21st century.

Career barriers are quite often created by women themselves. "Glass ceiling" is taken to be a phenomenon of nature. There is no way to fight natural phenomena: we live with them and take them for granted.

There may also be internal reasons that prevent the successful performance of women in science. We have problems working in a team: we do not support each other. Statistics show that in elections women vote against women candidates. Because mutual help and support are necessary elements of success in work, special training could help. We ourselves still have a lot to learn. We should take a critical look at ourselves and work on the problems we find inside.

GIFTED GIRLS

There are special schools in Russia that accept children gifted in physics and mathematics. Students must pass special exams to be admitted. The entry requirements for girls, however, are lower than for boys. Girls are accepted into these schools basically for psychological balance in the classroom. As a result, 80% of teacher attention is given to the boys, who have higher self-esteem and who are ready to participate in different competitive events. The boys are generally believed to have a higher IQ. In fact, girl students are often a burden to a teacher. The teacher contingent, therefore, should also be involved in discussions about improving the climate for women in physics. Teachers who work with girls should be encouraged to find ways to involve girls somehow.

I believe that some examples are helpful for involving gifted girls in the world of physics. At least in my life, an important part has been played by my professor of mathematics from St. Petersburg State University, Olga Ladizhenskaya. I always admired her at scientific seminars where she demonstrated a profound knowledge of material, put questions to a speaker, expressed her opinion on the subject. I learned from her that it is possible to reach a high professional level when you feel at home in your field. She has always been my role model. I have been trying to live up to her example all my life.

On the other hand, I know that nobody has to help you. Each person has interests and gifts. One should try, make mistakes, and never give up. If you find your mission, you shall learn that professional knowledge strengthens important elements of personality. You become internally stable and protected.

BALANCE OF FAMILY AND CAREER

I am quite confident that active occupation of a woman in science is compatible with organizing normal family life. A woman's mission is to be a mother to the same extent as a man's is to be a father. Children need both educated mothers and educated fathers. If a woman lags behind in active research it is not because she is less gifted.

She, the family woman, has to be perfectly organized. But it is the rare quality. The day consists of 24 hours. It is a lot of time, but the problem is how to organize it.

A national peculiarity in Russia is the institution of "babushka." It means that grandparents take an active part in bringing up grandchildren. They are a reliable help. But there is the reverse side of the coin: when parents grow older we take care of them and never think of it as a burden.

POSITIVE STEPS

Gender equity in science is a real problem to work on. The St. Petersburg Union of Women in Science was created two years ago. It consists of women physicists and engineers. In order to clarify the current state of the "gender equity" problem a special conference was convened in St. Petersburg that was supported by the Russian Academy of Sciences. Invited speakers were leading specialists in gender relations and the psychological, social, and biological issues of women. Speakers were men and women together. The conference turned out to be an interdisciplinary study of the gender equity problem. The subject of this presentation is a summary of what was discussed at this conference.

Activities such as conferences, lectures by successful women, writings about the contributions of women in different branches of physics, self-improvement of women, and mutual help comprise a short list of positive steps toward improving the climate for women in physics.

ACKNOWLEDGMENTS

The author is indebted to IUPAP and the organizing committee of this Conference for the invitation to participate and for financial support.

Women in Science: Personal Impressions

Catherine Cesarsky

European Southern Observatory

Catherine Cesarsky is director general of the European Southern Observatory in Garching, Germany. After receiving a diploma in physics from the University of Buenos Aires in 1965, she was a graduate student at Harvard University, where she holds a doctorate in astronomy. She then worked at the California Institute of Technology before returning to France in 1974. Dr. Cesarsky spent the major part of her career at the Commissariat à l'Energie Atomique (CEA). She was head of the Service d'Astrophysique from 1985 to 1993, and the Directeur des Sciences de la Matière at CEA, responsible for all activities in basic research in physics and chemistry, from 1994 to 1999. Until 1985 her research focused mainly on the origin of cosmic galactic radiation and the stray emission of gamma rays. She then became principal investigator of the infrared camera ISOCAM, which flew between 1995 and 1998 on the European satellite ISO. With ISOCAM results and now the European Southern Observatory's Very Large Telescope, she participates in advances in the understanding of the formation of stars and the evolution of galaxies.

Permit me first to say that I find it highly embarrassing to be standing here with the tag of "successful physicist." In the same vein, I am asked all too often to participate in "visiting committees" to laboratories or observatories, and when I accept and go, I invariably (anywhere in the world) find that women are too scarce and/or dissatisfied. Invariably also, at some point, everybody turns toward me and asks: "What should we do?" I make some general answer, but let me tell you, if I knew THE ANSWER, I would have publicized it years ago.

Even though I have given thought to these issues, my analysis does not compare with the excellent studies of the recent past. I can quote here, for astronomers, the studies published by the American Astronomical Society Committee on the status of women (Meg Urry, Lisa Frattare, et al.). And I have been tremendously impressed by the report from ETAN and Femmes et Sciences about "Science Policies in the European Union—Promoting Excellence Through Mainstreaming Gender Equality," which has been presented at this meeting by two of the authors, Teresa Rees and Claudine Hermann.

In the past 17 years, my only contribution to the questions of women in science, as director of large laboratories or of an international organization, has been to discreetly help to boost the careers and the self-confidence of capable women of all trades in my environment. Therefore, all I can bring to you here is impressions of what worked in my case, thinking that it could have for you, at least anecdotal value.

A MULTINATIONAL CAREER

I attended the French high school in Buenos Aires. All the good students moved on to attempt the "Grandes Ecoles" in France. Having little information, and being interested in science, I dreamt of what I had been told was the best school in Paris, Ecole Polytechnique. When an inspector came from France to survey our school, all the teachers selected me to be the student interviewed. He noticed, of course, and asked me at the end what my plans were. I talked about math, science, and Ecole Polytechnique. He laughed, and told me: "But there are no women at Ecole Polytechnique!" (This changed in 1972.) I decided then and there not to study in France.

I received an excellent training in Physics at Buenos Aires University, in conditions very similar to those presented at this conference by our colleagues from Argentina. I did not feel any discrimination as a woman, but years later I ran into the department chair in Geneva and he confided to me that he believed that physics was not for women, with very few exceptions (and I was not included in the exceptions).

As a graduate student at Harvard University in the United States, I also felt quite comfortable as a woman. Every year a woman (token woman?) was admitted in astronomy at Harvard, but in my year we were two. This made us both feel that we were really wanted, and gave us confidence to successfully achieve our degrees. Sandra Faber, my colleague, is now a very highly respected professor of astronomy at the University of California. She has, like me, raised two children, while consistently pursuing her career.

CP628, *Women in Physics: The IUPAP International Conference on Women in Physics,* edited by B. K. Hartline and D. Li

As a postdoc at the California Institute of Technology (Caltech), it was impossible, at least in 1971, not to feel awkward and uncomfortable as a woman. Consequently, with two other young woman scientists (one of them, Anneila Sargent from Scotland, whose name appeared at this meeting in the UK poster on distinguished women scientists), I created a support group for women undergraduates, a new feature then at Caltech. We obtained financial support from the Caltech Provost to conduct a survey about attitudes on women on campus, encompassing all the women, not just the scientists. We addressed the same questions to an equal number of men.

We publicized the results throughout the campus and had the impression that we had some real success in modifying the atmosphere on campus. Later we published an article with our findings jointly with a group of sociologists. In particular, we were struck by the fact that the ambitions and expectations of the women undergraduates were much lower than those of their male schoolmates. Was it realism (only one woman professor then in the University), a genuine lack of ambition, or rather the fear of success, of loss of femininity when stepping away from the traditional role?

I pursued my career in France, at the Commissariat à l'Energie Atomique (CEA), and there I always felt both accepted and valued, with courteousness and simplicity. I nevertheless felt like I was breaking the glass ceiling when I became the Director of Basic Research, with 3000 employees in my laboratories. But that nomination was well in line with a general, continuous trend toward the enhancement of women's careers at CEA.

I have reached a stage in my career where I do not think that it matters anymore whether I am a man or a woman. This is, of course, my own perception, which may or may not be warranted.

IMPORTANT ISSUES

Let me now summarize my personal views and experiences on a number of significant issues.

Husband: It is very important to choose well. The right husband, for a woman physicist, is not necessarily protective or even supportive (with some condescension!). He is an equal, respecting your choices and your career, and prepared to accept your independence—and even your success. My first fiancé, a brilliant mathematician, advised me to quit my studies when, in my third year at the university, I had some blues because for the first time I had encountered some difficulties. I instead quit him.

Children: Essential for me (not necessarily for everybody). I programmed my first child for the year of my Ph.D. But I lost that baby, and then, when given a postdoc at Caltech, it was intimated to me by the department head not to have babies while employed there. Instead, I became pregnant in my second month at Caltech, and was nevertheless able to continue my work there, with full support of my direct sponsors. Of course, no maternity leave, and a heavy load, but the science was great and the baby was adorable, so I was extremely happy. I had a second child later in France under easier circumstances. And I wrote one of my most important papers with this baby at my breast. I never interrupted my career.

Mentoring: I did not have one mentor; a number of people, including, of course, my thesis advisor, even though he was relatively remote (at Princeton), have taught me and helped me all along my career. But in reality, I have always been independent in my work. My most important mentor, between birth and age 13, was my brilliant eldest sister, Michelle, but then she left Argentina to study in France. I do not think that it is good for a woman scientist to depend too much on a mentor, especially if the mentor is a man.

Passion: I have always had a passion for science in general and for astrophysics in particular, and it grows with time.

Will power: Most likely my greatest asset. I always try to dominate my own character and impulses more than those of others.

Work power: High. It helps if you love your work, and if it does not force you to sacrifice other important things in life, such as family, friends, nature, art ... I have found a reasonable equilibrium or at least one suitable for me—but there has not been much leisure in my life!

Self-confidence: Not enough! I do not lack self-confidence toward other people, but when confronted with the riddles of nature: Who am I to solve them? As a result, much too often I do not publish my results, and/or do not push them to their ultimate conclusions. I nevertheless acquired a good reputation, in the first part of my career, as a theorist. Now I am more at ease, more confident in the validity of my results, using the fantastic new technologies, making observations, and interpreting them in a broad context.

Ambition: I certainly did not plan my career as it turned out. At first, I saw myself as a theorist, working alone in a corner with paper/pencil/computer; I soon realized the need to be aggressive in order to be heard or even noticed. And I adapted to that need. Once in Europe, I developed a strong wish to see France and Europe get ahead in science. It is because I thought I knew what to do, and because I was willing to take on responsibilities, that I ended up taking big management jobs. I started holding such positions when I was 42 years old. At the time, I was very surprised when I was offered the directorship of the large space laboratory where I was working; but I soon realized that this was a unique opportunity to obtain the necessary resources to properly develop the novel space instrument I had set my heart on. Nevertheless, my real ambition has always been to make discoveries. Today, I am quite satisfied with some of my results, but fear that, on that account, I did not completely fulfill my potential.

Special opportunities for women: Not a simple subject. While I see how these can help women's careers, I am ambivalent, because they can undervalue the recipient and undermine her self-confidence. (At this conference, however, Monika Ritsch-Marte told me: "I'd rather take the sneering with a job than the sneering without a job.")

Management style (Have I become a man?): I think there are differences between men and women managers, certainly between me and my male predecessors in my jobs as director. I, like them, believe in authority won through respect. But I listen more, delegate more, try to reach consensus whenever possible, and sometimes have a wish to nurture. I do not try to induce fear in my employees, even though I admit that it is sometimes a valid, and even successful management tool.

CONCLUSION: SAY NO

What is important, in the earlier years of the career of a woman scientist? To be able to say no, not indiscriminately, but at the right time, on the right occasions. No to second-rate situations, no to meddling in your personal life, no to always being second fiddle in a collaboration, no to a dominant partner. It still leaves you with many opportunities to say yes, and with the possibility of a happy, fulfilling life in science.

A Study on the Status
of Women Faculty in Science at MIT

Nancy Hopkins

Massachusetts Institute of Technology

Nancy Hopkins is the Amgen professor of molecular biology at the Massachusetts Institute of Technology (MIT). After receiving her Ph.D. from Harvard University in 1971, she was a postdoctoral fellow of James D. Watson and Robert Pollack at the Cold Spring Harbor Laboratory. With Watson and three others, she coauthored the fourth edition of the textbook, *The Molecular Biology of the Gene*. She was appointed chair of the first Committee on Women Faculty in the School of Science at MIT in 1995, and in 2000 she was appointed co-chair with Provost Robert Brown of the First Council on Faculty Diversity at MIT. She is a fellow of the American Academy of Arts and Sciences and a member of the Institute of Medicine of the National Academy of Sciences.

I am a molecular biologist and a professor at the Massachusetts Institute of Technology in the United States. Three years ago, in March 1999, MIT published the results of a study on the status of women faculty in its School of Science. My intention today is to review briefly what the study found, and then to tell you how MIT has responded to the findings. In particular I want to talk about some very remarkable progress for women faculty as a result of this study. However, last night one of your colleagues told me that, in addition, I should tell you a personal story of the MIT report. So, with apologies to several of you who have already heard this story, I will do so.

I would like to note from the start that when the MIT report was first published, it resonated with professional women in America. Not just scientists, but doctors, lawyers, women in business and government. Many wrote to tell us that our story was also their story. It seems to me from this meeting about women physicists in many countries around the world, that the MIT story was perhaps a very common worldwide story about the place of women in society and in the professions at the start of this millennium.

A PERSONAL VIEW OF MIT'S REPORT
ON THE STATUS OF WOMEN FACULTY IN SCIENCE

I joined the MIT faculty 28 years ago. I believed that the powerful Civil Rights movement that had taken place in America, and the Affirmative Action laws that followed, had solved gender discrimination. I did not expect to encounter it in my lifetime. I thought the reason there were so few women scientists at Harvard, where I was a student, or MIT, where I became a faculty member, was because women had children, and male scientists worked long hours. For women the two occupations seemed incompatible.

I found out I was wrong. I found out that gender discrimination still exists. It was a great surprise. It took me 15 years to figure it out. I figured it out by watching how other women were treated relative to men of comparable ability and accomplishment. It took me a long time to figure things out because there were so few women to observe. But after 15 years, I knew that men and women who were equal in accomplishment were not valued equally in our system. At first the women I watched were older than I. I was so glad I had come along later and thus could escape the unequal treatment I saw them experience. So I was very surprised when after some years I saw that women younger than I were not treated equally either. I wondered why I was the one exception!

One day, an administrator told me I was underpaid. I and other women in my department got an average raise of 20%. Despite this, it still did not occur to me that I was being treated unfairly. I was certainly very unhappy, but I thought it was my own fault. I thought my unhappiness was due largely to the ferocious competitiveness of many

CP628, *Women in Physics: The IUPAP International Conference on Women in Physics,* edited by B. K. Hartline and D. Li
© 2002 American Institute of Physics 0-7354-0074-1/02/$19.00

men in science, their inability to credit less aggressive people, particularly women, for their accomplishments, and my inability and unwillingness to behave like these men. Today, when I look back, I see that I was in denial. Denial is a wonderful thing! Denial protected me from the painful truth that I was not different from these other women. That I too was discriminated against.

But after about 20 years, several events happened that finally stripped the denial from even my eyes. For example, I needed a very small amount of lab space. I had a smaller lab than starting assistant professors and I had never asked for anything before. But when I asked for this tiny additional space, I could not get it. One day a woman who washed glassware for the labs said, "Nancy, how come these men have so much and you have so little!" It was that obvious, the difference. I struggled for more than 10 months to try to get this tiny amount of space. It was the accumulation of many incidents like this that finally enlightened me once and for all. These and what I had seen over 20 years in science. One day I knew I could no longer be a scientist unless things changed. I was very depressed for several days. But then, fortunately, one day I got angry. *Very* angry.

I decided that I would try to fix my problem. I would work my way up through the MIT administration until someone listened to me. Very soon I had worked my way right up to the president. I sat down to write him a letter: "Dear President, you may not know this, but there is discrimination in your institution. You should fix it." The letter was pretty strong so I decided to show it to another woman professor and ask her to edit out any comments that might upset the president too much.

I chose a female professor I scarcely knew, a highly successful woman scientist who I admired from afar. I asked her to edit my letter. We were sitting in a small café in Kendall Square near MIT. This was the hardest moment for me. I had never spoken to her or to other female colleagues about this issue. Why is this, you might wonder?

I realize now that many women never speak up because they think, as I had thought, that if you say you are discriminated against its like saying you aren't good enough. I felt that she would think badly of me. I imagined that she probably believed, as I had, that if you are good enough you can make it on your own—even in the face of discrimination. But I had found out this is not true. Even if she lost her respect for me, I would still be right. And I needed her help at this moment to make sure I wrote the most effective letter to the president.

I watched her read my letter. Her face did not change as she read. She did not appear to be changing her opinion of me as she read. Instead, when she finished reading the letter, she laid it down on the table between us and said, "I'd like to sign this letter. And I'd like to go with you to see the president of MIT. I have thought for a long time that senior women faculty here are not treated equally."

That was the moment that changed my life forever, and returned me to science. And that was the moment the MIT story began. It was not me. It was them. It was true. Women were not treated equally. It wasn't because they weren't good enough. It was because they were not perceived as equal. In that moment I think we both intuitively knew the power of what we had discovered. One woman alone, complaining, is just that, a difficult woman who can be dismissed. But two tenured women was a force to be reckoned with. We were a force because Civil Rights and Affirmative Action had put power into our hands.

We looked at each other and said, "You don't suppose there could be others?" We decided to make a list of the tenured women faculty in the six departments of science at MIT so we could poll them. That was when we made the startling discovery that there were only 15 tenured women faculty vs. 194 tenured male faculty in the six departments of science at MIT.

The very small number made it easy to poll the women. The first ones we approached responded with comments such as, "The same thing happened to me. Do you have anything I could sign?" They too had figured it out. These women scientists, who I had read about in the newspaper because they were always winning awards and being elected to the National Academy of Sciences. They agreed. When we finally located all of the women, about 10 agreed and could finish each other's sentences. Among the others there were a range of opinions: "It never happened to me but I know it happens to other women." "It happened to me some time ago, but not recently, but I want to join for the sake of my students."

In the end, all but one of the tenured women decided to band together and ask MIT to do something about the problem. It was the solidarity of these extraordinary women that made things happen. One woman alone could not be heard. But together their power was enormous. *The power was in the group!*

The women discussed ways in which to try to tackle our collective problem. We saw the issue as one of trying to help the administration understand the problem. Being scientists, we thought we should collect data. So we decided to ask the administration to let us form a committee that would (1) interview the women faculty and collect their stories, and (2) collect data in part to investigate incidents they reported wherever possible, and also to review equity issues such as salary.

At MIT the women were very lucky. We soon got the support of both then Dean of Science, Robert Birgeneau (a physicist) and President Charles Vest. They realized that you do not get near-unanimous agreement from a group of such highly successful and respected faculty unless there is a serious issue. They agreed to let us form a committee and perform the study we had proposed, even though the request met with considerable opposition at first from many other male faculty.

FINDINGS OF THE COMMITTEE ON THE STATUS OF WOMEN FACULTY IN SCIENCE

The committee, which I chaired, and which was made up of tenured male and female faculty, made the following important findings:

1. Young women, like older women before them, joined the faculty believing that gender discrimination is a thing of the past. They believed that only the greater demands of family that often fall to women will cause their professional lives to differ from those of their male colleagues.

2. Some time after tenure, however, many women faculty gradually came to feel marginalized. They saw that that their male peers had taken up the powerful positions in the department, while no women had held such positions. They often found themselves working harder and harder to achieve their success while the men seem to be working less to achieve comparable or even greater success. The men were often out starting companies in addition. Most men had families, more than half the women did not.

3. The committee found that there was good reason for the women faculty's perceptions. Indeed, the men often had more—not just in power, but in tangible resources and compensations. While each issue or reward might be small, one could see how they could add up over a career to produce the significant difference in status the women felt, and indeed often truly had. Most damaging perhaps was the exclusion of women from professional activities—from powerful decision-making positions and committees in the department, from important meetings inside and outside the Institute, from inclusion as founders of companies with male colleagues, from lucrative boards, etc., etc., etc. In many departments the women were all but invisible to their male colleagues. And, indeed, in number the women were close to invisible.

4. The percent of female faculty in the School of Science (8%) had not changed for at least 10 and probably 20 years. There was no indication it would change soon.

Clearly, by listening to the women, by putting their stories on the table, and by collecting data, the committee had succeeded in making the problems women faculty experienced understandable to the administration, and to the male faculty (themselves departmental administrators) who had served on the committee. The results were profound. First, the Dean, given documentation, corrected inequities immediately on a case-by-case basis. Although these were small things, the corrections helped enormously to restore a sense of fairness. In addition, MIT could begin to try to address the issues in a more systematic and systemic way. Furthermore, the media attention the report soon received led to outcomes that none of the original participants could have imagined—both inside and outside MIT.

PROGRESS FOR WOMEN FACULTY AT MIT SINCE 1999

MIT took three major steps in response to the School of Science report: First, the president asked that equity be monitored throughout MIT so that any inequities be identified and corrected at once, now and in the future. He asked that all the facts be on the table. The provost asked that committees like that in the School of Science be established in all five schools of MIT. This was done by the deans of Engineering, Architecture and Planning, the Sloan School of Management, and the School of Humanities, Arts and Social Sciences. Extensive reports were prepared by each of four committees of male and female faculty, each committee chaired by a tenured woman. The reports were released to the faculty on March 18, 2002. They are available at http://web.mit.edu/ faculty/reports/.

In addition to monitoring equity, the provost and the president took steps to address the issue systemically. They wished to ask why women, including women of color, as well as minority male faculty, all remain seriously underrepresented at MIT, as they are at essentially all leading research universities in the United States. To do this, they established a Council on Faculty Diversity. This council is co-chaired by the provost, me, and another professor who is African American. The council has three committees. One, the Quality of Life Committee chaired by Professor Lotte Bailyn, has prepared new family leave policies that have now been adopted by MIT in an effort to make it easier for women, and men, to balance family and work. A second committee, chaired by the dean of Engineering in collaboration with the provost and the other academic deans, has issued new hiring guidelines that will help to ensure that search committees have made their most effective effort to locate outstanding women and minority faculty candidates. The third committee will study pipeline issues, particularly for minority candidates, who often leave the academic profession before receiving their Ph.D.

The third important response of the administration was to aggressively recruit women to the academic administration at all levels. Whereas there were no women in these roles in science and engineering in 1994, today there are about a dozen.

Most women faculty at MIT recognize how fortunate we have been. Women at other institutions often tell us they still struggle to get their administrations to acknowledge these problems. I would attribute the success of what happened at MIT to several factors, but particularly the following:

1. A partnership of committed leadership with committed tenured women faculty.
2. The collaboration and mutual support of the tenured women faculty, first within each school and now across the five schools of MIT.
3. The willingness of women faculty and administrators to devote much time to this work, the willingness of administrators to reduce teaching loads for some women faculty who performed this work, and financial support from the Ford Foundation and Atlantic Philanthropic Service Company to help women faculty who perform this work. I would particularly acknowledge the support of President Vest and Provost Robert Brown, and the remarkable women colleagues who have worked closely together for some years now. I am particularly grateful to my wonderful colleagues in the School of Science and my close working group colleagues Professors Gibson (Engineering), Bailyn (Sloan School), and Hammonds (Science, Technology, and Society; and Director of the Center for the Study of Diversity in Science, Technology, and Medicine).

BUT THERE IS MORE TO DO!

Despite what many of us view as remarkable progress for women faculty at MIT, we know that much remains to be done. Given the small numbers of women in some fields, and the slow rate of faculty turnover, many women faculty in some fields of science and engineering will not have many, if any, close female colleagues within their departments and fields during their careers. How does one prevent the marginalization of these women? We can monitor equity and make sure they are equally treated, but how do we ensure that colleagues include them in important professional interactions? This is a challenge for the future and one that we are working on now with the chair of the faculty, Professor Steve Graves, and individual department heads. In the end, it probably requires greater awareness and understanding of these issues among all of our faculty.

IMPACT OF THE MIT REPORT ON WOMEN SCIENTISTS OUTSIDE MIT

The publication of the "Report on the Status of Women Faculty in Science at MIT," and its almost accidental release to the press, resulted in a remarkable public response. This came from women who had experienced the same problems but had not been heard, and from many who had been unable to hear them until the president of MIT heard the tenured women in the School of Science. In his comments that accompanied their report, the president wrote: "I have always believed that contemporary gender discrimination within universities is part reality and part perception. True, but I now understand that reality is by far the greater part of the balance."

The story received front-page coverage in the *Boston Globe* and the *New York Times*. We were deluged by e-mail from women at other institutions telling us of similar problems. I was invited to the White House where the President and First Lady of the United States thanked MIT and the women faculty. I was asked to give many talks on this topic at universities around the country. Wherever I went I met women grappling with the same issues we had identified. Given this response, the president of MIT called a meeting of the presidents of nine research universities to share experiences. This powerful group made a commitment to address bias on their own campuses, along with the underrepresentation of women and minorities in science and engineering, and to meet again to review their progress. A number of these and other universities have performed, or are performing, equity analyses similar to those performed at MIT.

Although the tenured women faculty in science at MIT had set about merely to remove obstacles to their research and teaching and to improve the situation for their students, it seems apparent that what they were dealing with are manifestations of broad societal issues involving the status of women and their role in the scientific and other professions. It will be interesting to look back in 20 years and see whether the efforts of the women faculty at MIT contributed to making scientific careers more accessible to a larger number of young women who share their passion for science.

Being a Woman Physicist: An Indian Perspective

Rohini M. Godbole

Centre for Theoretical Studies, Indian Institute of Science

Rohini Godbole is a phenomenologist with a special interest in high-energy particle colliders. She chaired the Center for Theoretical Studies of the Indian Institute of Science in Bangalore from 1997 to 2002. She has published more than 140 publications and guided the research of several graduate students. A manager of international collaborative research projects, Dr. Godbole is one of two women-physicist fellows of the Indian Academy of Sciences. She is a member of the National Board of Higher Education's committee to promote excellence in mathematics among women in India, and serves on several other national committees, including the expert committee for the CERN-India Project of the Large Hadron Collider and the planning committee for the Schools in Theoretical High Energy Physics Program of SERC (Science and Engineering Research Council). She is an associate editor of *Pramana* and the *Indian Journal of Physics*, sits on the editorial board of the interdisciplinary journal Current Science, and is referee for various national and international journals. She received her M.Sc. in physics from the Indian Institute of Technology in Bombay and her Ph.D. in physics from the State University of New York at Stony Brook.

INTRODUCTION

I focus in this talk on the following three issues in the context of Indian women physicists: (1) learning physics, (2) teaching physics, and (3) doing physics. I would like to present the thesis that at least in urban India, among those to whom good education is available, there is no substantive gender bias in the first two cases and the opportunities for women to take part in learning and teaching physics have steadily improved over the past decades. It is the third issue, doing physics, for which the situation is far from satisfactory.

Given the fact that easy availability of even primary education for women became a reality in India only close to the end of 19th century, the progress with respect to the first two issues gives me heart that the situation can improve in the third aspect as well, provided appropriate steps are taken. My discussions of these issues will include references to my own experiences, first while training to be a theoretical particle physicist and then while being one, spanning a total of 30-odd years, trying to identify what helped and what hindered this journey. I will conclude by telling of some nationwide initiatives, just in the budding stages, to improve the participation of women in science.

In India a student typically goes through 11 or 12 years of schooling (primary, secondary, and higher secondary), followed by three years of college toward a B.S., two more years, normally at a state-funded university, toward an M.S., and then joins a research program for a Ph.D. In some cases the B.S. and/or M.S. degrees are obtained at one of the eight Indian Institutes of Technology (IITs) or my own Institution, the Indian Institute of Science (IISc), which are the elite institutions of higher education in India. Most of the students who opt for professional courses such as engineering and medical do not then join these colleges. While the bulk of the research funding and activity in India happens in the research institutions, many of the good universities have active research groups focusing on somewhat smaller areas of research. Almost all the universities in India are state funded.

My world line has more or less followed the above-described general pattern. I studied in an all-girls high school for 11 years in my mother tongue, Marathi. I studied for my B.S. in a local college in my home town. This was followed by two years of a master's program in one of the IITs. I then worked for my Ph.D. at an American university. My post-Ph.D. time has been spent, almost in equal proportions, at a state university and the two most

CP628, *Women in Physics: The IUPAP International Conference on Women in Physics,* edited by B. K. Hartline and D. Li
© 2002 American Institute of Physics 0-7354-0074-1/02/$19.00

elite science institutions in India: the Tata Institute of Fundamental Research (TIFR) and the IISc. I began as a postdoctoral fellow at TIFR and am now a professor and Chairperson of the Centre for Theoretical Studies at the IISc. I have taught (and continue to teach) physics at various levels and am an associate editor of the *Pramana - Journal of Physics*. I have been involved in funding committees. Thus, I have experienced a broad spectrum of the Indian physics scene and should be able to communicate those experiences to you.

LEARNING AND TEACHING PHYSICS IN INDIA

I grew up in a typical middle class urban family, in a part of India which has always been very progressive about women's education/emancipation. One of the first Indian women to become a medical doctor in 1884 was from this state. The school where I studied was the third school for women to be founded (in 1885) in the state and most probably among the first 10 in the whole of India. All the women in my immediate family, including both my grandmothers, were well educated for their time, which meant until the eighth grade. My mother's education until the second year of college was supported by her mother, even in adverse economic conditions, and she was married at the age of 21, rather late compared with the prevailing practice then. My mother went back to university after three children and 15 years of marriage to earn her master's in linguistics and continue a B.S. in education. This background explains the full support and encouragement I received, and continue to receive today, from my family, and which has been very crucial for me.

Having said these good things, let me point out the problems too. While mine was one of the best schools, because it was a girls' school it was thought that we could begin studying science only in the eighth grade; however, science was a compulsory subject for some of the special scholarship examinations, held statewide to identify bright students in the seventh grade. I recall studying for the exam (in 1962) on my own and succeeding. To do justice to the principal of the school, the policy was changed after the disconnect was pointed out.

The science training in my school was generally not good. One woman math teacher, noticing this, directed interested girl students to a special training program in science. Interestingly, my mother was not able to study mathematics beyond the sixth grade because there were no women mathematics teachers and a man could not teach in a girl's high school in those days. In today's schools in urban India (many of which are coeducational), many of the science and mathematics teachers are women. Teaching is one job that many women feel they can handle well without compromising their commitment to their families. As a result, most often women with good educational records and abilities opt to teach. In one of the prestigious colleges offering a B.S. in education, more women than men have opted for special training in science and mathematics teaching for the past five years.

In India there are no social barriers to the science (particularly mathematics and physics) education of girl students. The fact that the majority of teachers are women, even for math and physics, definitely provides the girl student with confidence that she can study these subjects. Of course, one could also interpret, and with some justification, that the dominance of women in the teaching profession is the result of the field not being a very attractive profession to men anymore. Whatever the reason may be, the larger number of women teachers at the high-school level certainly means that learning physics and math is not perceived as either unusual or undesirable for high-school girls, as sometimes happens in some of the western countries. The high-school teachers with whom I've talked say they do not notice any substantive gender dependence in the interest of students in mathematics and physics. However, they do report a lesser degree of initiative and a higher degree of timidity in girl students.

At the college level also, the fraction of women students learning science is reasonable, close to 30–40%. Even in families that do not want to support a career in science for their daughter, people are happy if she takes up science, because it is seen as being more respectable and therefore helpful in finding a good match! (Here, of course, no preference is given for physics.) The percentage of women enrolling in colleges in physics has increased steadily from about 20% in 1969 to 30% today. Biological and chemical sciences have even higher female enrollments.

The substantial enrollment of women in physics programs continues into the master's level. (Up through the master's level in my own experience, no eyebrows were raised at my choice of careers.) However, it must be added here that the enrollment of women in somewhat higher-profile master's programs at the IITs or even at my own institute, where very often a move to a different city is involved, is much lower, at $\leq 20\%$. Again, in colleges and universities women form a large fraction of the teaching community. But not so in the more prestigious teaching institutes like the IITs or my own institution, where in physics we are just three women faculty members of a total of about 38. I believe this is at least partly a reflection of hiring bias. Women prefer to have these teaching jobs for the same reasons as outlined above for schools. By and large, the interest and ability of women teaching physics in colleges is distinctly superior because for women it is their first choice of career, unlike a large fraction of the men who teach in colleges. I can say this from my experience teaching refresher courses for college physics teachers.

If there is no shortage of role models for women physics students through the master's level, the situation starts changing at the doctorate level. The drop in women enrollment from the master's programs to the Ph.D. programs is substantial. I see the reasons as being (1) the possibility of having to leave home/family and relocate; (2) pressure from the family not to be "overqualified" lest finding a good match will be problematic; and (3) the length of time to complete Ph.D. and the consequent delay in starting a family. In my own case, my near-relations had tried to point out the second "problem" to my parents, who thankfully did not see it as a problem. The third reason, perceived incompatibility with starting a family, is something that will find echoes in my further discussions of physics as a career choice for women. Until recently, most of the fellowships available in India for Ph.D. work did not have defined guidelines about maternity leave. I was involved in the discussions that formulated some of these, and acutely felt the need for higher involvement of women in such policy bodies.

I should add here that a considerable fraction of the university faculty in India tend to regard the job as mainly a teaching job, with active research being optional. The implication for the current discussion is that a university job is very often "teaching physics" rather than also "doing physics." The treatment given to women faculty, say in distribution of teaching duties, has not always been free of gender bias. Again, to give a personal example, I was the first woman faculty at the physics department of my university. The then-head of the department was doubtful about my ability to teach the Mathematical Methods course, in spite of my being a rather successful theoretical particle physicist with a very strong scholastic record (I had always graduated at the top of my class) and a Ph.D. from a U.S. university. The most charitable comment I can make is that this was due to the department head's unfamiliarity with women teaching physics. This brings me to another of my beliefs, which I will state even though it is a cliche: the only answer to such blatant discrimination is to increase the number of women teaching at universities even more. In the 20-odd years since I had this experience, there has been a steady increase in the numbers, and a consequent *small* change in the attitudes.

DOING PHYSICS IN INDIA

This brings me to the most important issue, *doing* physics in the Indian environment: conducting and supervising research, writing and overseeing grants, and teaching. As we have seen, there is a transition loss at every level, but the two largest are from M.S. to Ph.D. programs and, particularly disturbing, from Ph.D. program to an active research career and a full-fledged faculty job that allows one to pursue her research. The percentage of young women faculty hired in physics is certainly not commensurate with that of women graduating with Ph.D.'s in physics. The most obvious reason is that in a two-career family, regardless of whether both are in physics/science, it is most often the woman's career that receives lower priority. Many women with Ph.D.'s in physics opt for jobs at government laboratories as technical staff due to the more fixed time requirements of such jobs. Knowing that capable young women often opt for these jobs, their senior male colleagues tend to suggest such softer options to them.

But even when this is not the case, the attitude of the senior members of the scientific community has been in the past very patronizing and presumptuous. I have sat on selection committees where I have had to hear that X will not be interested in a job because her husband and child live in some other town. Needless to say, the same concern does not arise while discussing the case of a male applicant. I witnessed people objecting to selection of a student for a graduate fellowship at my university, by saying that she is sure to get married and leave. Suggestions that we invite a woman colleague for a workshop have been objected to by questioning her level of participation in the activity because she will be accompanied by a child, in spite of the fact that the child was to spend the day at the day care. Such attitudes can only change when the numbers of women involved in the selection process increase.

The selection processes for various entry-level jobs or even postdoctoral positions do not take into account the time that a woman might have spent in starting a family. My suggestion that this point should be given consideration while comparing the publication profiles of two prospective applicants came as a surprise to some of my committee comembers. The issue needs to become a part of the thinking employed by the selectors. Certainly the first thing that needs to be done is to substantially increase the induction of women physicists into entry-level positions. It is clear that the same will then be reflected in substantive female representation in the various decision processes.

I was the first woman postdoctoral fellow to be employed in the theoretical physics group in the Tata Institute of Fundamental Research, the premier research Institute in India. After two years and two papers in *Physical Review Letters* in as many months, a senior colleague told me that my best bet was to get a teaching job in a women's undergraduate college in a nonurban area. How many different job levels he thought I was unsuitable for! Fortunately not all my colleagues felt this way, but the attitude reflected in the comment cannot be neglected, and I dare say that it had something to do with his perception about what a woman can do. Again, something that will change only with higher and higher participation by women in doing research. Certainly I do not notice this attitude among my younger colleagues. So again I am hopeful on that account. I will also be unfair to many of my senior and contemporary colleagues if I do not mention that their behavior to me has been quite free of any gender bias.

Another issue is whether there is discrimination in terms of being invited to give talks at professional meetings or other recognition by peers such as awards or election to the fellowships of the academy. The share of women physicists in India receiving fellowships and awards is minimal; for example, the percentage of women fellows in physics for the Indian Academy of Sciences (Indian National Science Academy) is 1.1% (0.9%), compared with an overall 4.07% (2.78%) for all disciplines. But the number of women holding jobs where one has a substantial amount of time and resources for doing research is already so small that these kinds of statistics do not mean much. Because many factors affect peer recognition, it is difficult to make a definitive statement about the effect of gender bias in these cases. Personal experiences of an obviously junior but male collaborator being invited to talk on a joint work cannot automatically be interpreted as gender bias, though the possibility cannot be dismissed, either.

On the very positive side, personal experiences, mine as well as those of my women colleagues, indicate that instances such as above, at least in the context of invited talks, do not happen these days—at least not in the small group of theoretical physicists with which I deal. I may add here that in other disciplines such as chemistry, biology, and medicine, where the number of active women scientists has been higher, high-profile positions such as presidencies of societies, top government positions, and membership of influential committees are occupied by women. About 18% of the fellows in medicine at the Indian Academy of Sciences are women. In physics, however, women of the same generation and similar academic stature have not necessarily gotten their due. Higher representation for women at the grass-roots level in research positions will help correct this imbalance.

In the earlier years of women doing science in India, most often a woman ended up giving a secondary position to her career. It seems to be still a somewhat sad fact of life in India that one needs a rather big dose of luck, along with a supportive partner and family, to have a successful career doing physics and at the same time a successful, complete family life. In most cases the success in career has come at the cost of commuting marriages, deferred child bearing, and in some cases families broken essentially due to the enormous strains that the simultaneous handling of two full-fledged careers and family bring to life. Personally, I went through a commuting married life for about 12 years, commuting across the continents, and thought in my not-so-youthful innocence, that these things can work out.

Another interesting aspect of the Indian psyche is, of course, that home management is perceived to be entirely female responsibility. Hence, if a male colleague is single the reaction is, "Poor fellow has to look after the house too!" But a single woman physicist "can perform better and devote more time to physics because she does not have to care for a home and family." Clearly, the norms of the society here have a huge gender bias. The guilt complex that mothers feel at not having enough time to spend with their children is accentuated by such societal thinking. It is also amusing to note that on the one hand a woman's possible commitment to the family and children is seen as a potential obstacle to her 100% commitment to her job, whereas a man taking time off for children's examinations is to be praised. The concept that either the man or the woman in the family has to find the time for the children and that it is the private decision of the family how this work should be shared, is simply not appreciated.

It is necessary to discuss these problems in different forums in the society and to try to change societal attitudes. In India there are few examples of women physicists who have come back successfully to physics after taking time off for child bearing. In the past few years, age limits have been relaxed for women scientists for various postdoctoral fellowships available in India. More such positive and forward-looking steps need to be taken.

HOPES FOR THE FUTURE

Fortunately, over the past few years there has been slow realization of the special efforts needed for improving the participation of women in science. The age limits for some of the special fellowships for young scientists, set up to encourage research in basic sciences, have been relaxed for women. This has been useful. I know of at least one woman (she had a postdoctoral position with me) who has benefited from the increased age limit and has been able to make a successful transition to active physics research, after a self-imposed child-bearing break, rather smoothly.

The National Board for Higher Mathematics is in the process of setting up lecture programs by active mathematician women to promote mathematics among girl children. The Board is also facilitating acquisition of a Ph.D. in mathematics by awarding special prestigious fellowships to women students, with special provisions for a break for child bearing, if necessary. Plans for creating special postdoctoral fellowships directed at giving the necessary cushion of time to settle two-career problems have also been explored. I am a member of the fellowship-creation committee, because theoretical physics and mathematics are perceived to have similar resource needs. This initiative has been followed by discussions with the presidents of the Indian Academy of Sciences and the National Board of Higher Mathematics, where the idea of forming an academy panel to look into the various academic issues of women scientists has been debated and should take shape soon. At the government level, the Department of Science and Technology is initiating programs with a view to increasing women's participation in science. On February 28, 2002, the Department announced 100 new fellowships meant only for women.

Of course, we have a long way to go. Most of the efforts at present have been directed at providing a good cushion of time to facilitate transition from a Ph.D. to a career in science. The issue of how to increase the number of women "doing" physics at the entry level needs further looking into. What is heartening is that the effort has begun.

Expand Research Space for Women in the Physical Sciences and Create a Better Tomorrow for Mankind

Zhili Chen

Minister of Education, People's Republic of China

Zhili Chen is mainland China's first female education minister. Her ambitious plans for education include reform of mainland China's higher education system and eliminating illiteracy among citizens aged 15 to 50. She studied physics at Fudan University in Shanghai and completed graduate studies at the Shanghai Institute of Silicates, a branch of the Chinese Academy of Sciences, in 1968. She then spent 15 years as a research fellow of that institute, specializing in ferroelectric and piezoelectric materials and devices. She was among the first group of modern Chinese scholars to visit the United States, as a visiting scholar to Pennsylvania State University from 1980 to 1982.

Science has always presented to people the excitement and stimulation that make us feel surging emotions and march forward in the face of difficulties. The 20th century, just passed, was an age witnessing leaps and bounds in the field of science and technology: from the maiden takeoff of an airplane to the landing of human beings on the moon, from the discovery of radioactive elements to the explosion of atom bombs and peaceful use of atomic energy, from the discovery of DNA to the birth of the sheep Dolly, from the invention of TV to the chess game between electronic computers and men. We owe all these spectacular achievements to the scientists and engineers who have been working wholeheartedly and walking strenuously ahead along the tortuous tracks.

We must never forget that under the brightly lit sky of science there are many outstanding women scientists. Without their contributions, the starry sky of science would have lost a great deal of its light, and the progress of mankind would have been slowed down considerably. The most brilliant and outstanding representative of women scientists is Mme Curie. Over 100 years ago, she discovered and extracted radium, and in 1903, she was conferred the Nobel Prize for Physics. She was the vanguard of women scientists who took the lead in researching physical science and made remarkable achievements in this field. What people respect most in Mme Curie is her diligence and persistence for her cause and her noble spirit entirely free from fame and gain.

Mme Chien-Shiung Wu, an American woman physicist of Chinese origin, is another representative worth mentioning in this field. It is known to all that ChenNing Yang and TsungDao Lee won the Nobel Prize for Physics for their theory of the "Non-Conservation of Parity." It was Chien-Shiung Wu, however, who first obtained the conclusive experimental results to prove their theory. It was these experiments that also established her historic position in the realm of physics. It is due to the persistent and indomitable spirit, keen observation and understanding that women exclusively possess that has helped them enjoy their unique strength in the field of physics research.

WOMEN IN PHYSICS IN MAINLAND CHINA

The Chinese Government has always upheld the equality between men and women, so that women's status has continued to rise in mainland China's political, economic, cultural, scientific, and technological life. It has also attached great importance to training women scientists and technologists, strove to improve their working and living conditions, and encouraged and supported them in their scientific research. Ever since the founding of the People's Republic of China, to be a physicist has been the ambition of many a young woman. Thanks to the encouragement and support of the government, the proportion of women in the field of physics has kept going up. A large number of women have entered the most advanced high-tech fields of high-energy physics, microelectron technology, and satellite launching, and scored remarkable achievements. Like their male colleagues, they have reaped breakthrough harvests of research and made indelible contributions to the development of mainland

CP628, *Women in Physics: The IUPAP International Conference on Women in Physics,* edited by B. K. Hartline and D. Li
© 2002 American Institute of Physics 0-7354-0074-1/02/$19.00

China's science and technology. In the Department of Mathematics and Physics of the Chinese Academy of Sciences, there are six women academicians.

Women have always been an important force in the field of physics in China. I would now like to mention in particular my teacher, the late professor Xie Xide, who, with her outstanding achievements in the field of solid-state physics, became well-known in the international community of the physical sciences. During her presidency of Fudan University, a famous higher learning institution in Shanghai, her able leadership helped this university leap onto a new stage and enter a fresh phase of development. Her characteristic charm became a spiritual force, educating and inspiring people to work hard for the shared lofty cause of humankind.

In mainland China, there are also great numbers of women who are engaged in teaching physics. Among more than 77,800 physics teachers of urban secondary schools across the country, over 30,600 are women, making up 39.3% of the total. Moreover, we have attached great importance to training women specialists in physics.

Women have made outstanding contributions to the development of the physical sciences, which have offered us wide space and theater for our development. Full of self-confidence, women physics researchers have worked harmoniously with their male colleagues.

The rigorous training and creative thinking required in the physical sciences has enabled professional women, who are already good at thinking both by instinct and in images, to enjoy the advantages of both logical and abstract thinking. The hard work required in the study of and research into physics has cultivated in them determination and persistence. All this has created a strong foundation for many of them to carry out further research in other sciences, and go in for social activities and public affairs. Some outstanding women in the fields of astronomy, automation, and computers studied physics during their university days. I feel very strongly about this experience myself. As a woman minister of education, I am proud that I was myself long engaged in research in the physical sciences, and I indeed benefited tremendously from this experience.

Due to the development and opening-up of our society since the reform and open-door policy, the level of education of Chinese girls has been raised by a big margin. More and more young women can enjoy post-university education. Girl postgraduates have reached 33% of the total postgraduate population. The options for women to seek their own development have been much broadened. However, importance must likewise be attached to the fact that the proportion of women engaged in the study of and research into the physical sciences has gone down. Take, for example, enrollment in the Department of Physics of Beijing University. The proportion of girl students in this department was 12.7% in the 1950s, 20.2% in the 1960s, up to 39.5% in the 1970s, down to 15.9% in the 1980s, and 4.6% in the 1990s. Fortunately, due to the expansion of education, the total number of girl students in this department has not been markedly reduced.

We have already taken note of this trend, and are adopting steps to encourage more young women to devote themselves to scientific research, especially in the physical sciences. The status of women in the political, economic, educational, cultural, scientific, and technological fields should continue to be raised, and the social and cultural environment for women's development should be optimized. The government of mainland China has also strengthened its efforts to compile comprehensive national statistics of women's development by establishing a women's databank and a gender statistics index, and by the collection, sorting-out, feedback, and exchange of information concerning women, including data on women devoted to research in the natural sciences. All of this has offered the government a sound basis on which to forecast women's development trends.

RECOMMENDATIONS

In order to attract more women to study physics and devote themselves to research in the physical sciences, I believe it is necessary to take the following measures:

1. Departments of physics and graduate schools should provide more chances to admit good girl students who are keen on studying physics.
2. Girl students of secondary schools should be invited to participate in the activities around small inventions and creations as well as other research work in the physical sciences. Competition programs exclusively for secondary-school girls should also be worked out.
3. Young women who specialized in physics should take advantage of their physics background by doing scientific research in much broader fields, and should be introduced to and encouraged to conduct research work in cross-subjects and other disciplines.
4. Women professors of physics from higher learning institutions and senior fellows pursuing research in the physical sciences should be mobilized and encouraged to care for and support the training of young women engaged in teaching and research in the physical sciences.

Since the mid-1990s, the Ministry of Education has set up in various universities 84 national bases for training talents in the basic subjects of science, including the physical sciences, so as to exclusively train gifted young people, both undergraduates and postgraduates, to conduct basic research. At present, a large number of outstanding graduates from secondary schools have been attracted to study and do research in basic subjects, and among them, the proportion of girl students has grown gradually. They are mainland China's prospective successors for basic research in the natural sciences in the years ahead.

I sincerely wish all women devoted to physics the greatest of success!

Papers by Country

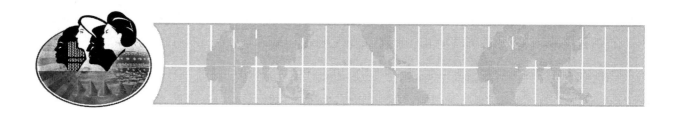

Posters from each country team lined the walls outside the plenary lecture room, providing a rich background for networking and information exchange throughout the conference. These posters can be viewed on the Conference Web Site: http://www.if.ufrgs.br/~barbosa/conference.html.

Women in Physics in Albania

Antoneta Deda

University of Tirana

In Albania, the number of women with careers in physics is very low. The majority of women who graduate with a degree in physics from the University of Tirana choose to teach physics in a middle or high school rather than pursue a career in research, which is a long, difficult road with many obstacles. The decision to teach is influenced by parents and husbands, who do not encourage careers for women in difficult fields such as physics because they believe the women would be unable to perform their family duties, which are considered more important.

One must remember that Albanians have lived for more than 50 years under a communist dictatorship that is very severe compared with those of other eastern countries. Albania's family life has been extremely hard, both economically and spiritually. Having patriarchal mentalities, Albanian men do little in the way of household chores and so are much more free to pursue careers. Complex disciplines such as physics, math, and chemistry have been monopolized by men. That is why it is necessary to increase the participation of women in physics in our country. Increasing the participation of women will help in the emancipation of women in work, society, and family. We must prove that women are capable of having successful careers in the sciences. This, above all, will help change the male opinion that women cannot succeed in the difficult areas of life.

A few of us, with much difficulty but also with much courage, have had modest, successful careers in physics. Albanian women who have pursued the different areas of physics take great pleasure in their work. I cite my own experience:

- It has been a great pleasure to me that students I have taught work all over Albania, and some of them, who have successful careers in physics, now come to consult with me about their problems.
- Being a professor in physics, I have been able to coach my children in their math, chemistry, and physics schoolwork. This is probably a higher priority for me than for many other parents, and has also given me great delight.
- I have had the opportunity to enjoy enriching scientific environments by participating in national and international conferences and exchanging experiences with colleagues.
- I would also like to emphasize my greatest pleasure, that of getting the best results of my career after a difficult, but beautiful job.

Unfortunately, there is no official staff to work on improving the working conditions of women in physics in Albania. A few people, especially women nominated to government positions, have had the opportunity to do something in this regard, but efforts have been spontaneous and the results have been disregarded. Our country does not offer many opportunities in scientific research. Everything takes a great deal of time and effort and you have to do it on your own.

Most Albanian women who have pursued a career in physics have worked in teaching and applied science. They have not made exciting discoveries, except what they have done on their dissertations, which has been useful and interesting, especially in teaching and applied physics. Although we have faced many constraints—lack of laboratories and equipment—we have been successful in hydrometeorology, and in processing materials such as copper, aluminum, and iron, which are the treasure of our mountains. Some of us are engaged in microelemental analysis of metals, agricultural products, and water, and some have careers in medicine.

Women who pursue scientific research in Albania are faced with, and have overcome, severe barriers:

- The mentality of parents who hesitate to send their daughters to school to study physics.
- Men's patriarchal thinking, which often includes not wanting their wives to have a career.
- Hardships that come about in marriages from the wife's involvement in science or other time-consuming work, when housework and children's education is considered the highest priority.

CP628, *Women in Physics: The IUPAP International Conference on Women in Physics,* edited by B. K. Hartline and D. Li
© 2002 American Institute of Physics 0-7354-0074-1/02/$19.00

- The small number of laboratories at schools, universities, and research centers, which makes the physicist's job more difficult, especially women physicists who have extra family commitments.

To remove the above barriers to women physicists, we would like to suggest the following ways to help Albania prosper:

1. Increase government investment in physics.
2. Encourage nongovernment organizations, who have much more money at their disposal than our government, to (a) include physics projects on their lists of priorities in order to raise teaching and research levels and (b) increase awareness through media, advertising, publications, etc., designed to bring about changes in attitude toward women's power and skills in physics.
3. Remove the gender bias in education in order to attract women and girls to the field of physics.
4. Pass laws requiring that women job applicants be given preference over male candidates when they have the same qualifications.

Albania is overwhelmed by many problems. During the last decade, very little was done to improve the job contracts of women in physics. Indeed, not much was done in physics that was relevant to the development of Albania.

This conference to increase women's participation in physics is a great undertaking and we are pleased to participate. Since our experience in this field is not considerable, this conference will help us to learn and have new experiences, hoping that in the future the social and economic situation in Albania will improve.

Women in Physics in Argentina

Silvina Ponce Dawson[1] and Karen Hallberg[2]

[1]Universidad de Buenos Aires, [2]Comision de Energia Atomica

PERCEPTIONS OF PHYSICS

Three Nobel Prizes in Science have been awarded to Argentines. Still, Argentine society does not have a strong interest in scientific issues, especially those related to the natural sciences. For example, of the three major national newspapers, only one publishes a weekly science supplement, *Suplemento Futuro* in the newspaper *Página 12*, whereas all three have weekly literature supplements.

In Argentina there is not a strong connection between scientific research, which is mostly carried out at public institutions, and the development of new technologies. Private industries prefer to buy imported technology rather than to develop their own. Multinational companies import technology developed elsewhere. Although most governments declare that scientific research is a powerful tool for the country's development, the support of scientific institutions is often perceived as an expense rather than an investment. This situation gets worse during periods of recession, like the current one.

Public high school science education is very poor. Most students perceive that to pursue a career in the natural sciences, especially in physics, would be too difficult. One is not very likely to find a job as a physicist outside a public institution, where salaries for research scientists are usually very low. Thus, there are few incentives, other than one's own vocation, to follow a career in physics.

PHYSICS STUDIES AT THE UNIVERSITIES

Argentina is a country of only 36 million people, in which the gap between the richest and poorest sectors of the society has increased dramatically over the last 10 years, and where about 40% of the population lives below the poverty line. Therefore, the number of people that study physics is a small fraction of an already small total universe (you can consult the numbers at http://www.df.uba.ar/~silvina/women/datos_mujeres.html). It is difficult to draw conclusions when statistics are calculated over a very small universe; in any case, we have observed some trends that we present next.

Physics is studied almost exclusively at public universities. The largest universities are located in big cities, which have a large proportion of middle class people and a rather dynamic societal organization, and where women constitute an important part of the working force. Most university students come from the same highly populated areas that contain the universities. The number of students who move to other places to follow a university career is not very large. The students who do move usually come from small towns.

At the largest public university, the University of Buenos Aires, there are about 100 beginning physics students every year. About 30% of these students are women. Between 30 and 40 students obtain their "Licenciado en Fìsica" degree (between a bachelor's and a master's degree at an American university) every year and, again, around 30% of them are women. Thus, the dropout level is the same among women and men. This same behavior is observed at other universities in Argentina.

The case of Instituto Balseiro, a unique institution in the country, is an interesting one. Located in a relatively small town, Bariloche, by the Patagonian Andes, the school requires that its students have already completed two years of related study at another institution. Students can follow a career in physics or nuclear engineering, after passing an entry exam with applicants from all over the country. Students at Instituto Balseiro live on campus and are supported by scholarships. The fraction of women that enter the Instituto Balseiro each year is around 8% of the total of accepted applicants, whereas the total proportion of female applicants is on average around 15%. The

CP628, *Women in Physics: The IUPAP International Conference on Women in Physics,* edited by B. K. Hartline and D. Li
© 2002 American Institute of Physics 0-7354-0074-1/02/$19.00

proportion of female students is thus smaller than at the major urban universities of Buenos Aires (30%), Córdoba (26%), and Rosario (37%). This might be related to the fact that women are more reluctant to move away from their families to follow their careers.

The proportion of female physics graduate students at both the University of Buenos Aires and Instituto Balseiro is slightly larger than it is for undergraduates (40% and 16%, respectively). We can think of three explanations for this behavior. The first is again related to relocating. Many students go abroad, mainly to the United States, for graduate school. We do not have hard numbers on this, but according to our own experience, it is more common for males to go abroad than for females. The second reason is related to the fact that salaries and scholarships for Ph.D. studies are very low—a CONICET (Consejo Nacional de Investigaciones Científicas y Técnicas, the main scientific research institution in Argentina) scholarship is around $700/month, while more than $1000 is necessary to cover the basic needs of a family—and women are already used to receiving lower salaries than men. Third, it seems easier for men than for women to find a job in the private sector once they have finished their undergraduate studies. In many cases, these jobs are computationally oriented, and it is much more common for men than women to be attracted to the computer industry.

CAREERS IN PHYSICS

The situation gets worse when one looks at the numbers of male and female professors or CONICET researchers: the proportion of women decreases considerably as the level of a position increases. For example, in the physics department of the School of Exact and Natural Sciences of the University of Buenos Aires, 34% of assistant professors are female, and just one full professor out of a total of 15 (7%) is female. The percentage of women at the largest physics departments is around 21% for assistant professors and 4.5% for full professors.

At CONICET, around 43% of the physics researchers in the two lowest-level categories of positions are women. At the highest level, 1 of 14 physicists is a woman; at the second highest, 2 of 50 are. The proportion of female physicists on review bodies at CONICET is 12%. In 2000, 13% of the CONICET projects in physics had a female director.

The situation is different at the physics and the material science departments of the Atomic Energy Commission in Buenos Aires, where women comprise a fairly large percentage of the total number of employees. Furthermore, the physics department has had female directors for many years. One of them also became president of this institution and a member of its directorate.

In general, however, women's roles in physics are skewed toward the lower levels. It is unclear whether this is a consequence of recent increases in the percentage of women physicists or whether it will persist into the future. Our own experience seems to indicate that scientific career advancement is easier for men than for women. Advancement in academic scientific careers, such as in physics, requires a degree of mobility that, in the case of married people, needs family support. In Argentina the number of research physicists is not very high, so it is important for young researchers to spend some time working abroad. This is more difficult for married women than it is for married men, unless the couple shares similar career goals.

Couples in which both spouses are scientists are fairly common, and they typically look for postdoctoral positions or permanent jobs at the same place. When this is not the case, women are more likely to give up their own careers, at least temporarily, in order to accompany their husbands. Acting in this way undermines a woman's career, and it could be the reason behind the smaller number of women in higher-level positions of the scientific research community. Apparently many women give higher priority to their families, at a cost to their careers. This intuitive observation is supported by the fact that many women who have managed to achieve the highest positions are either single or married with no children.

Physics in Armenia

Inna Aznauryan

Yerevan Physics Institute

The development of physics in Armenia is attributed to A. Alikhanian, who in 1943 established a research station at Aragats Mountain to study cosmic rays. In 1967, his efforts brought about the opening of a 6-GeV electron-circle in Yerevan. Many well-known physicists from the former Soviet Union took part in this research.

In 1961 with Alikhanian's initiative and supervision the International School of Theoretical and Experimantal Physics was opened in Yerevan. This school attracted many famous physicists from all over the whole world to participate as lecturers.

All this contributed to the development of physics in Armenia, and now Armenia has five scientific research institutes in addition to the Yerevan State University's physics facility, where research is carried out in almost every field of physics. A large number of women physicists are involved in the work at these institutions, among them six doctors of science and 43 doctors of philosophy. (In 1996, the population of Armenia was 3.8 million.)

The women physicists of Armenia pursue theoretical and experimental research in the following fields:

High-energy and elementary particle physics	Cosmic-ray physics
Nuclear physics	Gravitation
Cosmology	Biophysics
Nonlinear optics	Laser physics
Superconductivity	Physics of crystals

Five women physicists working in solid-state physics are members of the International Organization of Crystallography (Oxford). I. Aznauryan is a theorist working in the field of QCD and the quark model. Collaborating with a group of coauthors, she created a relativistic quark model that is successfully used for the description of hadrons, particularly hadron form factors in the subasymptotic region of QCD. In the perturbative region of QCD, she obtained important results showing that the asymptotic regime of QCD is not yet achieved and should be expected at transfer momenta significantly larger than are now available. At present, Aznauryan closely collaborates with TJNAF (USA). She has confirmed that, based on the data obtained at TJNAF, the asymptotic regime of QCD will be achieved at $Q^2 >> 4 \text{GeV}^2$.

A. Danagulyan and N. Demeckina are performing nuclear research using photon beams at the Yerevan electron accelerator. For a long time, a team of physicists from Yerevan State University and Yerevan Physics Institute led by Danagulyan and Demeckina has been studying the characteristics of nuclear reactions by measuring the yields of residual nuclei. Analysis of the data obtained confirms the role of the different reaction mechanisms in spallation, fission, and fragmentation processes. The isotope effect was also observed on different isotope targets. At present the research of the team is continuing on the proton and heavy-ion beams in collaboration with physicists from JINR (Dubna, Russia). Based on these investigations, the team has also developed isotope production methods for medical treatment.

In addition to pursuing her research, Danagulyan lectures on the physics faculty of Yerevan State University and is very involved in the growth of young physicists.

I wish to mention T. Asatiani, a women physicist of the elder generation, who has worked with A. Alikhanian since 1943. She is now the member of National Academy of Armenia. In collaboration with Alikhanian, Asatiani discovered the narrow atmospheric showers conditioned by nuclear processes. On the basis of the synthesis of spark chambers and strong magnetic fields, Asatiani and Alikhanian discovered the exact method for measuring the momenta of charged particles.

At present the flow of young women into physics remains steady; however, they, as well as the young men and older physicists, are experiencing difficult times because of the dissolution of the Soviet Union, which has created a general worsening of the situation in the former Soviet republics and meager funding.

CP628, *Women in Physics: The IUPAP International Conference on Women in Physics,* edited by B. K. Hartline and D. Li
© 2002 American Institute of Physics 0-7354-0074-1/02/$19.00

The Status of Australian Women in Physics

Giuseppina (Pina) Dall'Armi-Stoks[1] and Manjula Devi Sharma[2]

[1]*Defence Science & Technology Organisation,* [2]*University of Sydney*

The ratio of women to men in physics in Australia is quite somber. If recent published surveys are any indication, the future of physics and the level of participation, especially of women, looks very bleak. In one recent survey it was reported that the proportion of females taking year-12 physics in Australian high schools has shown little variation over the past decade, the average being approximately 29%. In addition, the percentage of females who withdraw from year-12 high school physics has increased in the last decade from 69.8% to 78.5%.[1]

It is also of concern that the proportion of girls taking year-12 high school physics in Australia at the end of the last decade has kept steady at 29%, while that of girls taking year-12 high school chemistry in Australia was as high as 48.8% over the same period.[1] These statistics highlight that there is a long way ahead to ensure that the participation rates of women in physics are comparable to those of other science disciplines.

Unfortunately, the situation does not improve at tertiary education. It has been reported that from 1996 to 1999 the rate of Australian university students enrolled in physics was on the decline.[2] This is further compounded by the fact that the ratio of women to men was quite low; however, the female participation rate did increase from 15% in 1991 to 22% over the 3 years from 1996 to 1999.[2]

The proportion of females undertaking fourth-year physics (includes honours, diploma, and master's preliminary students) in Australian universities has increased from 16% in 1991 to 25% in 1999, which correlates with the gender balance of third-year physics students. Furthermore, the proportion of females undertaking higher degree studies in physics continues to increase steadily from 12% in 1991 to 19% in 1999, perhaps due to restructuring of physics courses at universities to make them more flexible and attractive to students.[2]

These sorts of statistics are encouraging. But they should not overshadow the critical issues that the overall trend of physics enrollment at Australian universities is on the decline and that the gender ratio in physics is still low for women, despite the overall level of student enrollment in universities having increased.[2] Unfortunately, physics has not been able to promote itself well enough to attract the attention of students. There is a real need to promote physics as an enabling discipline providing fundamental advances upon which science and technology thrive. It is necessary to take initiatives to provide physics and particularly physics to females, because women can make a significant contribution to the science. We need to balance the ratio of women to men in physics.

Furthermore, there is a shortage of qualified and enthusiastic physics teachers in high schools. This is expected to get worse as the current physics teachers are retiring and at times resigning dissatisfied and disillusioned with the education system.

One step taken in Australia to promote physics and women in physics has been to establish a Women in Physics (WiP) group. Australia's first WiP group was formed in Adelaide, South Australia (SA), in February 1992. The main objectives of the group are to:

- Facilitate networking and provide support for women in physics.
- Encourage and support girls interested in physical sciences.
- Increase public awareness of women in physics.
- Oversee policy statements related to physics.

Currently the SA WiP group has up to 60 members, including students, postgraduates, teachers, and scientists in universities, government, and industry. To facilitate networking, the SA WiP group has regular support group and social/dinner meetings with presentations by special guests or members. At the commencement of the SA WiP, the group took steps to publicize and promote its existence around the state and around the nation. Articles were placed in *The Australian and New Zealand Physicist*[3] (now known as *The Physicist*), the SASTA (South Australian Science Teacher's Association) newsletter, and the WISNET (Women In Science Enquiry Network) newsletter. The group also took steps to liaise with the SA Women in Engineering group, and initiated the following:

CP628, *Women in Physics: The IUPAP International Conference on Women in Physics,* edited by B. K. Hartline and D. Li
© 2002 American Institute of Physics 0-7354-0074-1/02/$19.00

- The International Women in Physics Lecture Tour
- A series of Year 10 Girls Physics Workshops
- Visits to high schools and teachers' conferences
- Creation of the Clare Corani Memorial Physics Awards for the top second-year female physics student of each South Australian university

The National Women in Physics group was formed in 1994, with the SA branch taking the lead in establishing the group. The national WiP is a cognate group within the Australian Institute of Physics. The WiP has a biennial meeting as part of the Australian Institute of Physics' Congress. The current strategies and initiatives of national WiP group include the following:

- Administrate the International Women in Physics Lecture Tour begun by the SA WiP.
- Develop a website and set up procedures and guidelines.
- Maintain representation on the Australian Institute of Physics Executive Committee.
- Affiliate and liase with other science professional organizations, including the Australian Institute of Physics and Women in Engineering.
- Contribute to national science policies.

A WiP group has a leading role to play in promoting physics and achieving a more equal gender ratio in the physics discipline. However, it does need more support from governments and technology industries. Governments and technology industries working with and supporting physics professional groups such as WiP can put strategies and programs in place that address the critical issues preventing students, especially females, from being attracted to physics.

One strategy is to establish a program(s) to ensure that secondary schools and universities attract qualified and enthusiastic physics teachers and academics. Programs should also be in place to ensure that the educators and students, especially females, are supported, encouraged, and made to feel confident and enthusiastic about pursuing physics.

Another strategy is to determine the long-term skills and knowledge requirements of Australia's technology industries and then put in place programs to educate students to meet these requirements. This is not a new concept to Australian universities, but it is critical that this concept is enhanced and is supported long-term to cope with the advancements of science, especially physics research and development.

By recognizing and understanding some of the fundamental reasons behind the problem are we in a better position to address the problem. Then, by having long-term strategic planning and programs in place to address the problem are we able to promote physics more effectively and attract more students, both male and female, into the physics discipline. Our ultimate aim is to see a thriving physics community with gender ratios comparable to those of other professions and a general public that is better educated about the role of physics in society and the world.

REFERENCES

1. J.R. DeLaeter and J. Dekkers, The Physicist 38 (July/August 2001).
2. J. DeLaeter, P. Jennings, and G. Putt, The Physicist 37 (January/February 2000).
3. P. Dall'Armi and A. Ralston, The Australian and New Zealand Physicist (June 1994).

Women in Physics in Austria

Claudia Ambrosch-Draxl[1], Monika Ritsch-Marte[2], and Kerstin Weinmeier[1]

[1]*Institut für Theoretische Physik, Universität Graz*
[2]*Institut für Medizinische Physik, Universität Innsbruck*

When addressing the topic of women in physics in Austria, the name of Lise Meitner (1877–1968), famous Austrian physicist, immediately comes to mind. Contrasting her life and scientific career with the life and career of a contemporary Austrian woman physicist named Marietta Blau (1894–1970), several conclusions can be drawn, and maybe some implications for the present can be identified.

Although there was apparently no significant interaction between the two persons, there are several obvious parallels: both women grew up in prosperous and well-educated Jewish families, both studied physics in their hometown of Vienna. After receiving a Ph.D. in physics (L.M. in 1905 and M.B. in 1919) both decided to move to Berlin to continue their professional careers (L.M. started her long-lasting and fruitful collaboration with the chemist Otto Hahn, and M.B. took up jobs related to x-rays in medical physics).

Being Jewish Austrian citizens, both women had to flee from the Nazi regime after the Anschluss in 1938 and lived as immigrants in several countries.[1] Both women were nominated for the Nobel Prize by outstanding men[2]—and both failed to receive it, but were later honored by other awards. Both women never married and never founded a family of their own, and finally, both kept a life-long emotional bond to their home country with several personal visits, but refused to return to Austria for a scientific position.

However, while L.M. is remembered as a famous physicist and is a source of pride in Austria today, the name of M.B. is hardly recognized. Perhaps this is so because L.M. undoubtedly belonged to the innermost circle of physicists in her field and because her twice not being considered for the Nobel prize was very controversial in the scientific community, even at that time.

Even with such strong parallels between the two careers, it is very difficult to separate the issue of being a woman in physics from other, political issues, such as being a Jew. Nevertheless, L.M. herself believed some of the problems that she was facing were related to her gender.

Some of these parallels, however, such as the role of a sometimes merely "tolerated" exotic being, a life devoted entirely to physics, without a husband and children, an often tarrying acknowledgment of scientific contributions, are indeed symptomatic of the problems of many female physicists in the 20th century—and may even reflect the present day situation to a certain extent.

However, one factor that has left its mark on present-day Austria is easily identified: educated women who were forced to leave the country for good because of the war could not serve as role models for future generations. Even after almost 15 years of affirmative action, only very few academic positions in Austria are filled by women: presently there is one full professor in physics (in the Faculty of Medicine) and only a few associate professors in all of Austria. The overall percentage of female physicists in Austria is approximately 5%.

Missing scientific role models might be one of the reasons for the striking underrepresentation of female physicists in Austria, an assumption that is supported by the parallel situation in Germany. In other European countries the percentage of female physicists is not satisfactory, but it is still definitely higher than in Austria.

---

[1] L.M. spent 1938–1960 in Stockholm and her last years in Cambridge. M.B. moved to Mexico City. Albert Einstein tried to intervene on her behalf in order to improve her working conditions. In 1944 she went to New York and in 1960 she returned to Vienna because of health problems.

[2] Between 1940 and 1943 there were several nominations of L.M. together with Otto Hahn. In 1946 a nomination of L.M. together with her nephew Otto Robert Frisch for their explanation of the nuclear fission by Oskar Klein was supported by Niels Bohr. M.B. was twice nominated by Erwin Schrödinger for the development of photographic emulsions for studying nuclear processes.

CP628, *Women in Physics: The IUPAP International Conference on Women in Physics,* edited by B. K. Hartline and D. Li

Other important factors have deep social roots, such as the different education of girls and boys, whereby the definition of male and female roles hinders support for the technical and scientific skills of girls. Moreover, since childcare facilities, in particular for children below 3 years, are insufficient, challenging professions and family seem not to be reconcilable for many women.

The list of obstacles is long and cannot be discussed in a nutshell, but keeping all these facts in mind, there are many things that can be done to help change the situation. A federal law establishing the legal basis for equal treatment of both genders became effective in 1993. In addition to the mandate to evaluate the statistical patterns and to report on unequal opportunities for men and women, steps along the lines of an affirmative action policy were taken. In areas where women represent less than 40% of the employees, an increase of at least 20% has to be achieved within two years. In employee categories with a very pronounced underrepresentation of women (below 10%), the number of female employees is to be doubled within this period. Within each employee category, these measures stay effective until a ratio of 40:60% of the genders is reached. Furthermore, all qualified women who apply for a professorship have to be invited for an oral presentation.

Despite these seemingly stringent legal measures, the situation has not dramatically improved in the field of physics. One reason is that the 40% rate within an employee category has to be reached across the whole faculty or university. Thus it may happen that within a research field like physics this rate may never be reached if the rate of women in biology, for example, is much higher. But the highest hurdle is a lack of awareness about the problems female physicists have in reaching positions. As a consequence, there is no will to search for qualified women, although committees are obliged to do so. (Calls for applications have to be repeated if there are no female candidates.) There is strong evidence that women are only encouraged to apply for positions in order to fulfill the legal requirements, but that their applications are ignored. In that way individuals and committees think more of how to circumvent the law rather than to fulfill it.

Finally, we would like to state why we believe that female physicists are important. In addition to the argument of fairness and equal opportunities, a well-balanced distribution of women and men in any situation, such as within a working group, a profession, or a political formation, is a chance for the future of society. This also applies to the responsibilities of the physics community in connection with scientific research.

To reach a fairer situation, we have to support girls and women of all ages and categories, starting with early education and continuing up to providing grants, in order to have enough women available for higher positions.

Women Physicists in Belarus

Larissa Svirina[1], Galina Zalesskaya[2], Iryna Miadzvedz[3], and Iryna Lisovskaya[4]

[1]Institute of Physics NASB, [2]Institute of Molecular and Atomic Physics NASB,
[3]Belarussian State University, [4]Institute of Sociology NASB

HISTORY AND STRUCTURE OF PHYSICS EDUCATION

Since ancient times the women of Belarus have been characterized by their aspirations for education and enlightening activities. It is impossible to imagine the cultural and spiritual dawning of the Eastern Slavs without Efrossinya Polotskaya (c. 1112–1173), the daughter of Prince Georgiy Vseslavich of Polotsk. She founded convents and churches with the first scriptoria, which provided books for the schools she set up. Efrossinya Polotskaya was one of the first woman politicians, a diplomat and a peacemaker. Afterward many women of Belarus were successful in literature, botany, business, medicine, and other fields.

The revolution of 1917 played the key role in the formation of social and gender equality, making higher education free and available to all sections of the population and repealing the prohibition against "male" professions (such as physics) through a number of legislative acts. Today Belarus is a land of complete literacy, where 28% of the population has a higher education; half of these are women.

One milestone in the formation of science and education (including physics) in our country was the opening of the Belarussian State University in 1921, with its physico-mathematical department (later the departments of physics and radiophysics) where one could train to be a physicist. Another was the creation of the Academy of Sciences of Belarus in 1929, which became the center of scientific research and training of highly qualified specialists. Previously, one could obtain a higher education in natural sciences only in Vilnya (now Vilnyus), Moscow, or St. Petersburg. Today Belarus boasts of eight universities awarding a diploma in physics.

Every year in Belarus about 800 people, about 30% of them women, obtain a higher education diploma. This diploma corresponds to a Master's degree in many other countries: in addition to passing comprehensive exams, students must carry out scientific research, summarize it in a manuscript, and defend its results. A Candidate's degree in physics and mathematics (equivalent to a Ph.D.) can be obtained through postgraduate courses at the Academy of Sciences or at state universities. To obtain a Candidate's degree one has to carry out scientific research, publish the results in scientific journals (about 10 papers), prepare a dissertation, and defend it.

To be awarded a Doctor's degree in physics and mathematics, it is necessary to make original investigations confirmed in about 30 publications, write a dissertation (when working at the institute or attending doctoral courses), and defend it. Both Candidate's and Doctor's degrees are awarded by the specialized Council for Defense of Dissertations, whose decision is approved by the State Supreme Accreditation Committee of the Republic of Belarus. Professorial status is conferred on a doctor of physics and math who has a long record of teaching or who has prepared at least five candidates of science. Corresponding Member and Academician status is conferred on a Doctor of Physics and Math who has achieved extra scientific accomplishments. There are 13 Academicians and 14 Corresponding Members in Belarus, none of whom is a woman.

During 1997–2001, an average of 34 Candidate's degrees and 13 Doctor's degrees in physics and mathematics were defended each year; 7.2 (21%) and 0.6 (5%), respectively, were those of women.

WOMEN PHYSICISTS IN SCIENCE AND EDUCATION

Research in physics is carried out at 10 institutes of the National Academy of Sciences of Belarus (NASB), as well as in universities. The achievements of women in physics cover such fields as laser physics, nonlinear dynamics of laser and optical systems, relaxation processes of complex molecules in different phase states, atomic and molecular spectroscopy, biophysics, physics of the solid body, crystal optics, thermophysics, and nuclear physics. Women take an active part in international conferences and publish their works widely.

Today 1,490 researchers are engaged in physics at NASB institutes, of whom 504 (33%) are women. This is a fairly high index, although it is lower than total proportion of women researchers working at the NASB, which

CP628, *Women in Physics: The IUPAP International Conference on Women in Physics,* edited by B. K. Hartline and D. Li
© 2002 American Institute of Physics 0-7354-0074-1/02/$19.00

is 48%. Women account for 7% of Doctor's and 14.5% of Candidate's degree-holders at the NASB. Women physicists at the NASB quit their jobs only half as often and visit foreign countries twice as often (mainly to participate in conferences) as women of other specialties. From 1996 to 2000, 32 postgraduate students, of whom 5 (15.6%) were women, completed NASB postgraduate courses and presented their dissertations for Candidate's degrees.

Women under 30 constitute 10.7% of the total number of women researchers at the NASB. Women aged 30–39 are 17.3%, aged 40–49 are 28%, aged 50–59 are 27%, and over 60 are 6.5% of the total number of women researchers at the NASB. Such a distribution is also characteristic of men and, accordingly, to the whole physics department of the NASB. Thus, we can speak of a certain "aging" of science.

Physics is taught as a general education subject in 27 higher schools. The trend of the higher schools is represented by the number of women physicists working in the departments of physics of two state universities. Eighteen women (17%) work in physics at Belarussian State University (1 Doctor of Science and 11 Ph.D.'s) and 14 women (21%) at Gomel State University (1 Doctor of Science and 4 Ph.D.'s). Twenty-five women (28%) work in physics at Belarussian State Polytechnical Academy (12 Ph.D.'s), the largest in Belarus' technical higher-education establishment.

The most representative group of women physicists with a higher education in Belarus is teachers of physics in the secondary schools: they number over 4,000 (about 80%). Teachers are paid less than physicists who work in industry, and the prestige of the work is lower than in science—traditionally this social level is occupied by women. For a woman, a positive aspect of teaching is that she has more free time and can do a portion of the work at home. Because women teachers are graduates of the state universities, they are highly qualified. Precisely this factor is the reason, that even in recent years when the prestige of physics among the young has dropped, the level of education is still high. This enables graduates of the universities to work in various fields of the economy in either Belarus or abroad.

The number of women physicists in industry is relatively low because it also requires engineering training.

CONCLUSIONS

In our country, the problems of women in the field of physics have not been studied; we hope that the activities of our group will permit us to begin such investigations. Many Belarussian women have chosen physics for their profession. The chief problem of women physicists in Belarus, as in other countries, is the relatively slow promotion rate and the low probability of occupying a leadership position.

A woman's career in physics will be more successful in our country if the level of industrialization of housekeeping is raised. Because of the relatively low salaries, few families can afford the modern domestic appliances (vacuum cleaner, dishwasher, clothes washer, etc.) that facilitate housework.

Labor legislation of Belarus defends the rights of women. At the state enterprises there are more permanent positions than temporal ones. When a woman is gong to have a baby, she has two-month paid holiday before the birth and receives partially paid holiday for childcare for three years. Nevertheless, women physicists meet various sorts of discrimination, such as less equipment, smaller offices, lower salaries, and psychological pressure from male colleagues. A key measure toward a more comfortable existence of women in physics would be the legislative regulation of gender balance at the administration level of research institutions, universities, scientific councils, and financial agencies. However, in our country, as in some others, all leading positions, both in government (10% of Parliament is women) and in scientific and educational establishments, are occupied by men. So at present it is difficult to adopt such laws (officially there are not gender problems).

One way to improve the institutional climate and decrease the discrimination of women is to create non-state women's commissions and work groups, probably within the framework of Belarus Physical Society. Our participation in the present conference allows our team to speak of the problems of women physicists in Belarus, to learn about the lives and work of our foreign colleagues, and in particular, to avail ourselves of their experience gained in professional women's organizations, which will make it possible, after joint discussions, to work out a number of proposals for improving the position of women in physics.

SOURCES OF DATA

1. *Higher Education in Belarus, A Brief Guide*, Ministry of Education of the Republic of Belarus and National Institute of Higher School affiliated with Belarussian State University, Minsk, 1998.
2. Annual Reports of the Ministry of Education of the Republic of Belarus, Minsk, 1995–2000.
3. Attectaciya *1*, 29-31 (2001).
4. *Annual Report on Activity of the National Academy of Sciences of Belarus*, Minsk, 2000.
5. Personal correspondence.

Women in Physics in Belgium:
Still a Long Way From Achieving Gender Equality

Petra Rudolf[1], Christine Iserentant[2], Muriel Vander Donckt[3], Nathalie Balcaen[2], Peggy Fredrickx[4], Karen Janssens[4], and Griet Janssen[4]

[1]Facultés Universitaires Notre-Dame de la Paix and Vice-President of the Belgian Physical Society; [2]Universiteit Gent; [3]IBA, Louvain-la-Neuve; [4]Universiteit Antwerpen

UNIVERSITY STUDENTS

In Belgium the university student population is nearly equally split among men and women. Of the 131,415 students enrolled during academic year 1999/2000, 66,667 (50.7%) were female.[1] The 750 physics students represent just a tiny portion of this population, 0.57%; the field is clearly male dominated, since women comprise only 26% of the physics students.

Gender-specific statistics on the Belgian physics student population have been available only since 1995. They have shown that for the last five years the female proportion of physics students has always been between 24% and 26%. Among those who received their bachelor's or master's degrees in physics during academic year 1999/2000, the female percentage was the same, at 22%. The percentage is slightly higher, 24%, among the Ph.D. students who defended their theses that same year. These numbers are still far from 50%, and they indicate that the success rate of women in their bachelor's studies is slightly lower than that of men.

In interviews conducted with students in various universities, female physics students often denounced the lack of encouragement from their families. The prejudice that "physics is not for girls" seems widespread. We believe that this attitude leads to a lack of confidence, which is probably responsible for the slightly higher dropout rate of female students.

UNIVERSITY TEACHERS

If one looks at the totality of physicists teaching at Belgian universities, women make up 12.6%[1]; however, this includes not only tenured senior assistants, lecturers, associate professors, and ordinary professors, but also Ph.D. students and postdoctoral researchers who work part time as teaching assistants. Considering the various categories separately, we have complete data only for the Flemish universities: in physics and astronomy nine of the 141 tenured academic and research staff (6.44%), 25 of the 43 temporary assisting academic staff (58%), and 86 of the 268 temporary research staff (32%) are women.[2]

For the French-speaking universities, the numbers for teaching personnel are available only for science faculties as a whole. The percentages of women are 9.7% for tenured academic staff, 28% for the tenured assisting academic

---

[1]Numbers collected from the heads of physics departments of the 10 Belgian universities delivering bachelor's (kandidaat/candidature) and/or master's degrees (licentiaat/licence) in physics, from the Conseil des Recteurs des universités francophones de Belgique (CreF), and from the Vlaamse Interuniversitaire Raad (VLIR). Belgium keeps different statistics for the French-speaking and Flemish universities, and the career structure is different for the two.
[2]Data for 1999.

CP628, *Women in Physics: The IUPAP International Conference on Women in Physics,* edited by B. K. Hartline and D. Li

staff, and 42% for temporary assisting academic staff.[2] Among the physicists appointed for research positions in the French-speaking universities by the Belgian National Fund for Scientific Research (FNRS), the female percentage of tenured positions is far lower (10.9%) than for temporary positions (20.8%).[2] This shows that the "leaky pipeline" exists in Belgium for access to permanent academic and research positions.

Certainly, one has to take into account that 20 or 30 years ago the percentage of female students was lower than it is today, which obviously limited the percentage of female candidates.[3] Unfortunately, we don't have numbers that would allow us to evaluate this factor. However, the higher female percentage among the tenured assisting academic staff than among tenured lecturers, associate professors, and ordinary professors in French-speaking universities indicates that it is more difficult for women to reach the higher rungs of the career ladder.

Several countries have reported discrimination against women in the allocation of research grants—Belgium is fortunately not affected by this problem. In fact, if one looks at the numbers for 1998/1999 for the FNRS, the percentage of female grant applicants is always exactly the same as that of female grant recipients.[4]

ACCESS TO POWER

So far as the access of women to power structure is concerned, there is much room for improvement, as the following examples illustrate: On the board of directors of the FNRS, only three out of 23 are women. There are no women on the board of the Flanders Fund for Scientific Research (out of 26 members), nor on that of the Belgian Nuclear Research Centre. For the Interuniversity MicroElectronics Center (IMEC), Europe's largest independent research center in the field of microelectronics that is located in Flanders, only two out of 10 members of the board of directors are women. Among members of the scientific commissions of the FNRS who judge funding applications for physics (grants and fellowships), the female percentage is 10%. None of the heads of physics departments of Belgian universities is a woman, and one-third of the physics departments do not have a female professor. This obviously means that there is an acute lack of role models for young female physicists.

Very few universities have created a committee for equal opportunities, and the debate on this issue within academic institutions is just now getting a timid start. Networking and mentoring activities are still very scarce: the network of women in science in Belgium was born this year (http://bewise.naturalsciences.be) and the Belgian Physical Society has just started its working group.

RECOMMENDATIONS

We believe that in order to improve the percentage of women physicists in Belgium, two types of actions are necessary. First, one has to attract more female students and act against the prejudice that "Physics is not for girls." Second, those who finish a Ph.D. or a postdoctoral research appointment have to be prevented from dropping out from research and academia.

To attract more female students, it is necessary that successful women physicists become more visible to potential future colleagues through increased participation in the power structure. Without this, all personal efforts of women physicists to promote their field among girls (through visits to schools, TV programs, newspaper articles, etc.) lack credibility and cannot be successful. Reducing the dropout rate can be achieved by avoiding strongly competitive or aggressive environments, in which women often do not feel at ease, in the workplace.

Another discouraging factor that should be addressed is the lack of women in governing bodies. Additionally, practical aspects—such as measures to facilitate the reconciliation of men's and women's professional and family lives—play an equally important role in the decision of a woman to remain in the field of physics.

---

[3]In the French-speaking universities, equal numbers for male and female university students were reached only in 1992; in 1987/1988 the female percentage was still 43% (CreF).

[4]See 2001 ETAN Report on Women and Science: "Science Policies in the European Union: Promoting Excellence Through Mainstreaming Gender Equality," www.cordis.lu/improving/women/documents.

Low Participation of Girls in Science Programs in Botswana

Sisai Mpuchane

University of Botswana

Botswana is a landlocked country in southern Africa with a population of 1.7 million. It is also a member of the Southern African Development Community (SADC). When it gained independence from the British in 1966, it was the 25th least developed country; but through its fortuitous discovery of diamonds and the development of its beef industry, it has transformed itself into a democratic country with a vibrant economy. The gross domestic product has grown by 6% per annum and the government has given high priority to education in all of its development plans. The number of primary and secondary schools throughout the country has increased rapidly. Primary education is compulsory and by and large free.

A 1998 the Gender Empowerment Measure of the United Nations Development Program gave Botswana a ranking of 48 out of 174 countries, and a 1997 government survey showed that only 25.7% of administrators and managers were female (the female population is over 51%).

Participation levels of women in science-based careers are very disappointing. Several studies have been carried out to identify the cause for this discrepancy and while the government's education policies do not discriminate against sex, few women enter the science and technology realm. We continue to list the causes for this discrepancy—cultural beliefs, women's social obligations, male dominance, among them—without coming up with sound solutions. Clearly, something needs to be done to address the underrepresentation of women in scientific and technological activities.

Subjects taught in our primary schools normally include mathematics, science, social studies, Setswana, and English. Recent Ministry of Education data (2001) on primary school exit examinations reveal that primary school girls do in fact register for various subjects (Table 1).

Table 1. Primary School Exit Examinations Showing Grades A–C.

Subjects	Male Students	Female Students	Total
Mathematics	13,084	14,921	28,005
Science	13,634	12,878	26,512
Social Sciences	15,911	17,216	33,127
English	12,650	15,430	28,080
Setswana	14,093	17,123	31,216

In 2001, 40,213 candidates sat for the primary school exit examination: 20,742 females and 19,471 males. Students performed best overall in Setswana, with 86.5% obtaining grades of A–C. English (76%), social studies (74.8%), mathematics (67.4%), and science (64.4%) followed. Girls performed better than boys in all subjects in 2001.

In Botswana, the enrollment of girls in primary schools is higher than that of boys. Their performance is also better in all courses including mathematics and science and, given more encouragement, they could become good scientists. The quality of science education is important because it determines whether our nation will ever build a strong science culture. The participation of women as primary school science teachers is important because it exposes girls to the necessary role models at an early age.

There are also more girls in secondary schools in Botswana. There are more female teachers at the primary level than at the secondary level. In fact there are far fewer mathematics and science female staff than biology staff. As girls register for courses at the senior secondary level, they drop the pure sciences and opt for the integrated sciences. This creates problems for their admission into the science schools of tertiary-level colleges.

CP628, *Women in Physics: The IUPAP International Conference on Women in Physics,* edited by B. K. Hartline and D. Li
© 2002 American Institute of Physics 0-7354-0074-1/02/$19.00

A 1996 survey of women scientists in the SADC region revealed that there were 20% women in the biosciences, 14% in chemistry, and 4% each in physics and mathematics. Girls are therefore deprived of mentorship by female teachers at secondary schools. As a result there is a low enrollment of girls in chemistry and physics (mathematics is compulsory).

At the tertiary level, female participation in science and technology is also low. The University of Botswana, the only university in the country, has an enrollment of female students in the Faculty of Science of 28% and 12% in engineering and technology. In other faculties (humanities, education, and business), the participation levels are representative (>50% of total enrollment).

The statistics for academic staff in science at the University of Botswana (around 27%) clearly indicate a need to improve participation so that female students might be properly mentored by female role models. Botswana is not different from its SADC partners in this regard, as evidenced in a 1993 survey of gender representation on university staffs undertaken by the Third World Organization of Women in Science (TWOWS). Results are shown in Table 2.

Table 2. Staff by Gender and Position in SADC Countries, 1993.

Rank	Male (%)	Female (%)
Professors	12.93	0.71
Senior Lecturers	23.79	1.74
Lecturers	53.1	7.61

Concerns have been raised about various factors that disadvantage girls at schools. The dropout rate due to pregnancy, the higher incidence of HIV/AIDS among females (a 1999 UNAIDS report indicated Botswana had the highest incidence in the world, at 36.1%), strong sociocultural beliefs of male dominance, and general illiteracy.

Female university graduates in physics are few, as demonstrated by the output from our Faculty of Science for the past 5 years (Table 3):

Table 3. Female University Graduates in Physics at the University of Botswana, 1997–2001.

Year	Male	Female
1997/98	28	1
1998/99	43	10
1999/2000	33	5
2000/2001	19	2
2001	4	2

It is therefore very difficult to identify female physicists in the workplace because of the low number. The Faculty of Science at the University of Botswana also reflects this shortage of academic staff in physics (Table 4).

Table 4. Female Academic Staff in the University of Botswana Faculty of Science in 1998.

Department	Female Staff Establishment	% Female
Biology	8/24	33.3
Chemistry	4/24	16.7
Computer Science	3/19	15.8
Environmental Science	7/29	24
Geology	0/10	0
Mathematics	2/26	7.7
Physics	0/24	0

Botswana needs to explore various ways of addressing this discrepancy. Attempts are being made through creating some enabling policies, networking with various organizations (e.g., UNESCO, TWAS, TWOWS, NORAD, FEMSA, WISTAN, the Rockefeller Foundation, USAID, and FEMED), and through other government and nongovernment programs (mentorship workshops, science clinics, fellowships, and training).

The University of Botswana has just started a project that encourages secondary school girls to study science. This project includes science clinics, publications, a motivational video on women's participation in science and technology, conferences, and visits to schools by female scientists.

Women in Physics in Brazil

Marília Caldas[1] and Marcia C. Barbosa[2]

[1]Universidade de São Paulo, [2]Universidade Federal do Rio Grande do Sul

The scientific community in Brazil has been growing steadily in the last 50 years. Because the history of science in the country is relatively recent, there is no national media for dissemination of science in Brazil. The overall population thus has very little knowledge about the level of the scientific research that is being undertaken in the country, and science is seen by the common citizen as something mysterious, difficult, and usually done by "unsocial" people.

There is no difference in the way an elementary or high school teacher will treat a girl or a boy in math or physics classes in Brazil. However, a majority of the children will develop an aversion to the subject itself, and the prejudice against scientists has a stronger effect on girls who, in a male-dominated society, usually look for social acceptance. Women play a very important role in the educational structure in Brazil because the mother is more involved with the education of the children, and because most of the teachers at the elementary level are women. Therefore, if the way women view science and the scientific method is not reversed, we will not be able to change the way the overall population feels about science and research.

The underrepresentation of women in science in Brazil has other contributing factors as well:

- Access of women to education is recent. During the time Brazil was a colony of Portugal (until 1822) and for the first 50 years after its independence, most of the wealthy families educated their children in Portugal, where the first schools for girls were opened in 1815 and women were not allowed into universities until 1879. Sending girls to study outside Brazil was never a priority for Brazilian families, so women actually had access to higher education only when our first university was opened in 1912. Even then, they would prefer to go into the humanities. Today the proportion of female undergraduate students in physics is only about 25%

- Women face yet another difficulty in pursuing graduate and postdoctoral studies: reconciling career and family, particularly for young physicists. It is very important for a Brazilian physicist to spend at least one year in postdoctorate studies outside the country, which is often difficult, if not impossible, for a married woman, particularly one with small children. As a result, after graduating, women will decrease their participation in science. This can be gauged by the number of study fellowships granted for physics by our national agency, CNPq: we find that female participation starts at 25% for undergraduate grants, and drops to around 11% at the postdoctoral level.

- Physicists in Brazil basically work at universities. As researchers and faculty we have a double career. Researchers can achieve national recognition through a research fellowship from CNPq, which in turn has important effects on the everyday life of the scientist: getting new grants, participating in plans, reviewing, and so on.

- Fellowships are classified into six levels, and requirements for reaching the top level are not objectively stated; they are associated with "seniority" that does not have a clear definition. Without clear rules, there is plenty of room for prejudice. The physics committee is trying to have clear or at least more openly posted rules; however, right now the situation is of grave concern, with 17% female participation at the entry level dropping to 1% at the top. Faculty members are also classified in levels, and the rules are again not completely clear. Since there are few women at the top levels of the field, committees that make the decisions are male dominated.

- Very few women have top administrative positions in our grant agencies, so the rules and decisions are made from a male perspective.

How can this situation be reversed? We need to work on the issue from two fronts: by attracting girls to physics and by making it possible for those who are in the field to pursue successful careers. We propose these strategies:

1. Improve the teaching of science at all levels, making it clear that it is not a male activity.
2. Create working groups inside the societies and at the universities that will address issues specific to women and fight against the barriers.
3. Make sure that promotions at all levels of the physics career path—decisions about grants, fellowships, and so on—are based on clear rules.

The Status of Women in Physics in Bulgaria

Ana Proykova

University of Sofia

Scientific results are obviously independent of the gender of the researcher, but the methods chosen for problem solving and the style of presenting results at scientific forums are often gender-dependent. Usually, women are better speakers and explain the subjects more carefully. This may be a result of women's experience with children at home. While raising and educating the kids, the woman should bring the abstract knowledge of the world to the child's mind. Thus, women may better understand the difficulties that can arise from the special use of everyday words in a particular scientific field. This natural ability determines their career: they become teachers in physics, more often at high schools than at universities.

The choice of approach and methodology of investigation, if there are options, is based on an intuition distinct from the male's. A man asks a question like: Does the content of thought merely give us abstract and simplified "snapshots" of reality, or can it go further, somehow to grasp the very essence of the living moment that we sense in actual experience? Searching for the answer takes a lifetime. A woman usually considers problems that can be resolved in a limited time: a month, a year. Of course, there are exceptions on both sides.

The twentieth century opened the doors to the scientific temples for many women. This resulted in a recognized contribution of women to fields like biology, medical science, biophysics, and chemical physics. The number of women in "pure" physics was smaller than in these other disciplines. This is a general observation. Contemporary research is more interdisciplinary than field-limited; in this respect, the importance of women in physics-dependent fields will grow.

In Bulgaria the number of female and male students at the universities was kept the same for many years: admission was 50-50 during the communist regime. However, the number of female Ph.D. students was smaller than the number of male students. It is a matter of long discussion why the number of women in top positions was, and still is, negligibly small, although the number of women educated in physics is relatively high.

The working conditions for women in physics at home are not different from those for men. However, the expectations of the society are different: men are expected to be inventors and women are supposed to be performers. This can make women less motivated at work. The expectation of the society is only one aspect of an indissoluble process of interplay between a person and a society.

The 50-50 admission of girls and boys before 1990 is one of the reasons for the presently high number of women at the Bulgarian universities and in physics. Another reason is the incredibly low income in academia: men, who are still considered responsible for the family finances, escape to better-paying jobs in industry. Permanent positions and control of faculty activities, though ineffective, contribute to a noncreative environment. With the lack of financial support from the government, it is understandable that few achievements are encountered in physics, including from women, of course. Most contributions are made via international collaborations with Bulgarian physicists participating: nuclear spectroscopy of high-spins, thin-layer production and optical study, plasma physics, theoretical physics/group theory, statistical and computational physics.

Some of the women (not a complete list) who are still working and have contributed to nuclear physics are:

- Prof. Dr. Sci. Krasimira Marinova (University of Sofia)
- Assoc. Prof. Dr. A. Minkova (University of Sofia)
- Sen. Res. Dr. Dora Kutsarova (Institute of Nuclear Physics, Bulgarian Academy of Science)
- Sen. Res. Dr. Sci. Doriana Malinovska (solid-state physics; Bulgarian Academy of Science)
- Sen. Res. Dr. Diana Nesheva (solid-state physics; Bulgarian Academy of Science)
- Assoc. Prof. Dr. Krasimira Germanova (solid-state physics; University of Sofia)
- Assoc. Prof. Dr. Ivanka Jordanova (solid-state physics; University of Sofia)
- Prof. Dr. Sci. Antonia Shivarova (plasma physics; University of Sofia)
- Sen. Res. Dr. Ana Georgieva (theoretical physics; Bulgarian Academy of Science)
- Sen. Res. Dr. Boika Aneva (theoretical physics; Bulgarian Academy of Science)
- Assoc. Prof. Dr. Ana Proykova (theoretical physics; University of Sofia)

CP628, *Women in Physics: The IUPAP International Conference on Women in Physics,* edited by B. K. Hartline and D. Li
© 2002 American Institute of Physics 0-7354-0074-1/02/$19.00

Only one women (A. Shivarova) is a full professor at present, after the retirement of K. Marinova this year. The most important woman's name from our past is Elisaveta Karamihailova, who in 1947 established the Department of Atomic Physics at the University of Sofia after she returned to Bulgaria. She was one of the few females who worked at the Cavendish laboratory with Thomson and Bragg, after working at the Radium Institute in Vienna for several years.

Most of we women in physics have found that the only way to improve our research and to increase its impact is to work abroad or to pursue international collaborations while keeping our positions in Bulgaria. Honestly speaking, the professional situation is not very different for male physicists. However, the difference between men and women is that the men hold the leadership and determine the scientific policy. Men are on the committees electing heads of departments, and they choose men. A possible reason for the unfair selection is that women usually consider the family more important than their career. At this stage we encounter a typical "Catch 22" situation: the administration restrains women from obtaining higher positions because of their family duties, which are constantly increasing because their husbands are at the same time putting more and more effort into climbing their own career ladders. If a woman breaks the rules by reaching the highest professional position, she risks harming her family life.

What can be done? We should start at home by educating girls that science, and especially physics, is not limited to males. And we need to achieve a new perspective by mixing the male and female way of thinking. Then we should continue this process on a societal level. If society expects girls to contribute to the field, then girls will be motivated to compete. We currently see that many young girls are being attracted by fashion and the arts, which are the completely opposite science. This is wrong.

At home in Sofia, we organize meetings of physicists. More and more women group leaders attend and give talks about their own research. This is very encouraging. An important element within these workshops is the development of a new style (for Bulgaria) of research collaborations: short-term, project-specific collaborations of researchers belonging to different fields (e.g., physics, biology, chemistry). With optimal use of existing expertise about complex systems, these collaborations allow each participant to contribute at her best to a higher knowledge.

Women and Physics in Cameroon

Ndukong Tata Gerard[1] and Samba Odette Ngano[2]

[1]Institute of Physics, Cameroon; [2]University of Dschang

As many conscious societies strive to give all their citizens equal opportunities, the under-representation of women in physics is becoming a glaring aberration and an embarrassment to governments and education planners. The worry (especially for developing countries) stems from the realization that physics is at the root of all meaningful technological innovation and development. Physics teachers in Cameroon are constantly reminded that, "Physics teaching in the secondary school should have the triple role of preparing students for tertiary education, producing middle-level manpower and producing a science-literate population."[1]

The under-representation of women in physics means that Cameroon women will lose out in the competition for top jobs in a largely technology-based job market. A democratic society requires that its citizenry is capable of making informed choices. In a world where science-related issues are constantly taking center stage, Cameroon risks having an inept womenfolk who lost the opportunity to become knowledgeable in science-related issues. The global effort to empower women must start with ensuring that the educational system offers the girl child the same opportunities as the boy child.

FACTS OF THE MATTER

Of the six state universities in Cameroon, four have no female physics lecturers at all, and the other two have only one each: Dschang University has one female physics lecturer in a staff of 11 for the department and the University of Yaounde has one female lecturer in the school of Engineering. Of the 423 Physics teachers that have graduated from the Department of Physics of the College of Education, only 25 are women, a ratio of 1:17!

This under-representation is even more pronounced at the level of student enrollment. Of the 145 first-year students in the Physics Department of Dschang University, for example, only 10 are female, a ratio of 1:14.5. Of the 96 students in the second year, only 6 are female, a ratio of 1:16. Of the 80 students in the third year, only 7 are female, a ratio of 1:11.5. Statistics from the same university show that of the 155 graduates from the Physics Department since 1993, only 24 have been women, a ratio of 1:7.

A survey of the staff of most industries and engineering firms clearly shows that women do not yet occupy top technical posts. Where women are found at all, their role is usually of a secretarial nature. The under-representation found in higher education is clearly evident in engineering and industry.

The situation is not better at the secondary-school level. As Tima (2000) points out, in Anglophone classes (Cameroon is a bilingual country with French and English as the official languages), there were 312, 100, and 26 girls in forms 3, 5, and 6, respectively, as opposed to 295, 236, and 281 boys, respectively, in the same classes.[1] In Francophone schools, there were 218, 82, and 51 girls in "troisieme," "seconde," and "terminale," respectively, as opposed to 273, 216, and 200 boys, respectively, in the same classes. Of the 56 Francophone physics teachers that responded, only 4 were women. Of the 106 Anglophone respondents, only 3 were women. Incidentally, the best national performance for the year was by a girl, who had a grade of A in physics.

The above statistics clearly show that girls and women are grossly under-represented in physics and related disciplines at all levels of our educational system.

REASONS FOR THIS STATE OF AFFAIRS

There is no evidence of a gender discrepancy in the ability to learn physics. Many reasons have been advanced for the poor state of gender representation in physics and related disciplines: ignorance of the problem's existence, the dis-enabling sociocultural environment of the Cameroon girl child, curriculum content, instruction methodology, absence of a realistic policy for equal opportunities in education, absence of role models, and an extremely examination-conscious educational system. These issues cannot be analyzed in a short writeup of this nature.

WHAT IS BEING DONE ABOUT IT?

While very few people would doubt the commitment of the Cameroon government to empowering women, the Ministry of Women's Affairs has yet to initiate and push through relevant reforms in educational practice. While efforts such as that of the First Lady and the Cameroon Radio and Television Corporation must be credited for highlighting issues connected with the marginalization of women, the credit for practical effort at encouraging the girls in physics lies with educational nongovernmental organizations such as the Institute of Physics–Cameroon (IOPC), the Female Mathematics and Science Teachers Association (FEMSA), the North-West Association of Physics Teachers (NWAPT), the South-West Association of Physics Teachers (SWAPT), and the Mathematics Teachers Association (MTA). The greatest handicap for these associations is surely finances. With a little help, they could have a great impact on the inclusion of girls in physics.

REFERENCE

1. R. Tima, "Problems encountered in the teaching and learning of physics in secondary schools in Cameroon: their impact on the uptake of physics by girls," Journal of the Physics Teacher, NWAPT *11* (2000).

The Canadian Challenge:
Attracting and Retaining Women in Physics

Marie D'Iorio[1], Janis McKenna[2], Ann McMillan[3], and Eric Svensson[1]

[1]National Research Council of Canada, [2]University of British Columbia, [3]Environment Canada

Canada continues to face a challenge in attracting women and retaining them in physics-related positions. There are a large number of studies of women's representation in the workplace[1-4] and a number of recommendations have been provided, but there has been little enthusiasm to implement many of these recommendations in order to evoke real change. Canada has in place a number of programs, including paid maternity leave, affirmative action programs, and awareness programs of various kinds, but, nonetheless, the representation of women, especially in the sciences, remains discouragingly small. Reasons for this remain a matter of debate; factors clearly playing a role are the "two-body" problem arising in the career development of the two-scientist family, the lack of female role models to influence girls and young women in schools and universities, the continued perception among young people that science, and physics in particular, is "difficult" and not very interesting, and the fact that child care remains primarily the responsibility of the woman.

A decade ago, an international study on gender distribution in physics departments[5] showed that the representation of women in North American physics departments was of the order of 4%, compared with 23–47% in Western and Eastern Europe. In 1995, the Committee to Encourage Women in Physics of the Canadian Association of Physicists sponsored a survey of physics departments in Canadian colleges and universities to obtain data on women in physics. The findings showed that although women obtained 18% of the B.Sc. degrees in physics and 13% of the Ph.D. degrees, only 5% of faculty members and 2% of tenured faculty members were women.[6] At the time, 11% of faculty positions were tenure-stream positions and women held 28% of these positions.

The numbers gathered six years ago painted a rather bleak picture in which 80% of the 40 Canadian institutions that responded to the survey had either one or no woman on staff. The results of a new survey being conducted in 2001 will be presented in the poster of this Conference. The indication from a Statistics Canada Labour Force survey is that in 2000 only 2.8% of women worked in the natural sciences, engineering, and mathematics fields, compared with 1.8% in 1987.

A number of programs have been implemented to help improve the Canadian environment for women in physics. In 1997, the Canadian government announced that, through NSERC (the Natural Sciences and Engineering Research Council), it would fund five new Chairs for Women in Engineering and Science designed to encourage the participation of women. The chairholders, chosen from different Canadian regions and scientific disciplines, were specifically mandated to develop strategies to encourage female students in elementary and secondary schools to consider careers in science or engineering, as well as to sensitize faculties on how to improve and promote the integration of women students and professionals.

NSERC also instituted a University Faculty Awards Program for women who would like to remain in or return to a Canadian university research environment. The program offers five years of salary support at the assistant professor level, a guaranteed research grant, and a tenured or tenure-track position at the end. The goal of the University Faculty Awards Program is to decrease the underrepresentation of women and Aboriginal peoples in faculty positions in the natural sciences and engineering by encouraging Canadian universities to appoint very promising researchers in those groups to tenure-track or tenured positions in science and engineering.

An example of a program to promote scientific leadership among undergraduate female students is the Women in Engineering and Science Program launched by the National Research Council of Canada (NRC) in 1991. Each year this program provides 25 new fellowships allowing the best female undergraduates in science and engineering to work in NRC's laboratories for three consecutive summers. The program has been structured such that each student is provided with a mentor who can help in the selection of a summer research project and a project supervisor. The students are also required to develop their leadership and communications skills by introducing the program to first-year undergraduates at their respective universities. This program has helped introduce female undergraduate students to the challenges of a research career, and encouraged them to continue on to graduate studies. Undergraduate summer and co-op employment programs in university, industry, and government laboratories have also contributed by offering women research experiences in different settings.

One of the programs that provides role models to school children as well as professional development support to science teachers is the Let's Talk Science Program. This national award-winning program matches graduate student volunteers in science disciplines with elementary-school and high-school teachers in one-on-one science partnerships. These partnerships benefit the volunteers, who hone their science communication skills, the teachers, who can enhance the science learning experience in the classroom, and the students, who get to meet graduate students pursuing a career in science. Currently 14 program sites exist across Canada.

The goals of Let's Talk Science are to interest young people in science, improve the confidence and competence of science teachers, and encourage participants to continue in a lifelong journey of learning science. This program is especially important to young women whose interest in science needs to be captured early by appropriate role models. Prominent role models have been played by the Canadian female astronauts Roberta Bondar and Julie Payette, who have inspired young women to pursue undergraduate and graduate studies in science and engineering. Last year, Dr. Payette lent her name to a scholarship for male and female graduate students showing outstanding academic excellence, research ability and potential, and leadership and communication skills.

It is clear that much remains to be done to empower girls and young women to take up physics. While progress is being made, there is still a problem with women dropping out of physics programs at each level and there is a very low representation of women at the most senior levels in universities, industries, and government laboratories.

REFERENCES

1. "Beneath the Veneer," The Report of the Task Force on Barriers to Women in the Public Service (Canadian Government Publishing Center, 1990).
2. N. Morgan, "The Equality Game, Women in the Federal Public Service (1908–1987)" (Canadian Advisory Council on the Status of Women, 1988).
3. "Report of the National Advisory Board on Science and Technology," presented to the Prime Minister of Canada (Committee on the Participation of Women in Science and Technology, February 1988).
4. P.J. Caplan, *Lifting a Ton of Feathers, A Woman's Guide to Surviving in the Academic World* (University of Toronto Press, 1994).
5. W.J. Megaw, "Gender Distribution in the World's Physics Departments," paper prepared for Gender and Science and Technology 6, Melbourne, Australia, July14–18, 1991.
6. J. Lagowski and J. McKenna, Women in physics in Canada, Physics in Canada *52*, 106 (1996).

Women in Physics in Mainland China

Ling-An Wu

Chinese Academy of Sciences

Although it is well known that the magnetic compass, rockets and other technologies were first invented in China, physics only began to develop as a science here in the early 20th century. Despite the social and political upheavals of that century, physics education and research managed to survive and indeed, flourish. The Physical Society located in Beijing was established in 1932, even as the country was under the yoke of foreign invasion, and it was precisely during this period that a young Chinese woman, Chien-Shiung Wu, was receiving her college education. Wu was to become the first woman president of the American Physical Society.

Other women followed in Wu's footsteps, continuing their educations in the United States and Europe because most universities were disrupted by the war. Many returned and later played prominent roles in building the physics infrastructure after peace returned. The earlier pioneers were involved in nuclear physics, reflecting the interest of their times, but others turned to solid-state physics, including Xi-De Xie, who, in the mid-1980s, was president of Fudan University, one of the major universities of mainland China.

In the 1950s the universities underwent a major reshuffle with strong emphasis on science and technology, and mainland China embarked on an ambitious plan to train new generations of scientists and engineers to meet the demands of a modern socialist society. Women were further emancipated, enjoying equal rights and equal pay with men and special benefits such as maternity leave. Enrollment of women in schools increased at an unprecedented rate, especially in the sciences, and in certain fields such as medicine the enrollment of women rivaled that of men.

However, physics has only seen a moderate rise, and the reason is by no means simple. In certain normal universities the proportion of female to male students has been as high as one third, but in the top-ranked comprehensive and science and engineering universities the number has seldom exceeded 20%. It seems impossible to derive an explanation just by looking at the statistics, which are illustrated in Figures 1 and 2 for Beijing Normal University and Nanjing University.

We see that the ratio of women students as a function of year over the last decades appears to be totally chaotic. This is because the enrollment number is a variable of social, political, and economic factors, all of which have undergone, and are still going through, major changes in China.

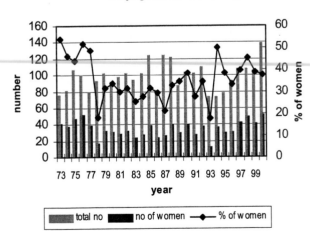

Physics Undergraduate Enrollment In Beijing Normal University

Figure 1

Formerly, the quota for physics departments in universities was entirely determined by the state, and enrollment was strictly based on examination performance. Then, during the latter years of the Cultural Revolution when university education resumed, the students were recruited in part by recommendation based on work performance. Since girls tended to be more docile than boys, they won more approval from the authorities concerned, which explains the sudden peak in the ratio of women in physics departments in those years. In the 1980s, national college entrance examinations were resumed, and after the "opening up" of mainland China to the outside world, external factors became a strong influence. The CUSPEA program initiated by Nobel prize winner T.D. Lee encouraged young students, boys and girls, to choose physics

CP628, *Women in Physics: The IUPAP International Conference on Women in Physics,* edited by B. K. Hartline and D. Li

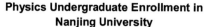

**Physics Undergraduate Enrollment in
Nanjing University**

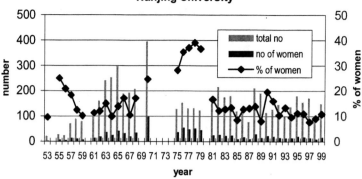

Figure 2

so that they could compete purely on the basis of scholarship to study for a Ph.D. in the United States. Participation in the International Physics Olympiad further enhanced this enthusiasm. But the ratio of women in physics has not dramatically increased, and in fact in some universities has decreased compared with pre-Cultural Revolution days. This may be explained in part by the emergence of new disciplines such as computer science and environmental science, as well as the revival of departments such as law, economics, and finance, which have attracted a large proportion of women.

As is commonly found, although the situation in mainland China is better than in many other countries, the ratio of women who remain in physics after college decreases with every step up in rank, and the ratio of women in the highest rank (equivalent of full professor) has never exceeded 10%. It is significant also that about 30% of the teachers of high school physics are female, and the total proportion of women in the physics faculty of some universities is now higher than 30%. However, the number of women in research institutes is rapidly going down, due to the impending retirements of a previous generation of women physicists with no new recruitment in sight. This phenomenon is evident from the statistics of the Institute of Physics of the Chinese Academy of Sciences (Figure 3). Recent figures from the National Natural Science Foundation of China show that the percentage of women awarded research grants was around 11.7% for the age bracket above 40 years but about 5.4% for women younger than 40.

The outlook for women in physics professions in mainland China is also disquieting for other reasons. Ostensibly there are still equal pay and equal opportunities for women, but many companies and institutions are now openly, though not legally, stating a preference for male employees. Moreover, there is now an official preference for young people when considering promotion or funding, which in fact implies discrimination because it makes it more difficult for women to return to a physics career after the years spent on child rearing and family duties. Many day care centers have been abolished for economic reasons, and the return of a certain degree of unemployment has even led to talk of letting the wife stay at home. Even male physicists have been heard to say there is no need to encourage women to take physics. This is particularly distressing when so many of them were taught physics by a female teacher in high school or college!

Institute of Physics Research Staff

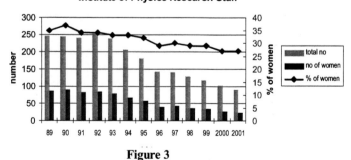

Figure 3

In the last two decades, mainland China has made impressive strides in modernizing its economy, on a solid foundation of science and technology built up over the years. In this, physics has and will always play an indispensable role. But in physics, women have been the unsung heroines. It is imperative that, with appropriate policies, women be allowed to give full play to their abilities in a career in physics, and thus make their maximum contribution to this supreme science.

The Status of Female Physicists in China-Taiwan: Balancing Family Life and Career

Jauyn Grace Lin[1], Ming-Fong Tai[2], Hue-Min Wu[3], and Yeong-Chuan Kao[1]

[1]*National Taiwan University*, [2]*Culture University*, [3]*National Chung Cheng University*

This paper reports on the status of female physicists on the island of Taiwan. As in other places in the world, female physicists are a minority there. We explore the probable causes for this phenomenon, focusing on the balance between family life and career for female physicists. Our findings apply equally as well to working women in other professions.

In February 2001, at the annual meeting of the Physical Society located in Taipei, we registered a Working Group on Women in Physics (WGWIP). This act coordinated with the inauguration of WGWIP in the International Union of Pure and Applied Physics (IUPAC) at the March 1999 General Assembly. The major functions of the WGWIP are to:

1. Offer consultation to female physics graduate students on issues of personal affairs and career difficulties;
2. Support female physicists trying to further their careers;
3. Improve the working environment for female physicists on the island of Taiwan; and
4. Participate in international affairs on related issues.

Based on recent statistics,[1] 3.8 million unemployed women between the ages of 15 and 64 live on Taiwan, and more than half of them are married. Among these unemployed married women, 40% said they quit their jobs willingly after getting married. The main reason they gave was their wish to take care of their children at home. These facts show that family duties may be more important than developing careers to women.

Nevertheless, these data also reflect the difficulties in balancing a family and career for these women. In general, a married woman is automatically given the job of home management. She will have to give up her profession unless she has enough energy and resources to conduct two jobs. The following information may help us to understand why most women on Taiwan cannot find the balance between family and career.

According to the latest public survey,[1] a full-time housewife spends 3.1 hours per day on housekeeping duties, and 2.2 hours on child care. If elders live with the family, a housewife is also responsible for accommodating their daily needs. A married and employed woman spends 2.8 hours per day caring for her children and 3.1 hours working in her house. These data demonstrate that a working woman holding a full-time job does not necessarily have a reduced workload at home.

An employed married man, on the other hand, spends 1.7 hours per day helping with child care and housekeeping. These data clearly indicate that a married woman has to work 4.2 hours a day more than a married man to keep both the family and her career going. Generally speaking, this would discourage women from having their own careers after marriage. How about women in physics? Are they better or worse off?

We first look into statistical data of men and women in the Physical Society located in Taipei. According to a report by Prof. S.N. Yang, the ratio of female to male physics graduate students is 16% and the ratio of female to male physicists working in academic or research institutes is 10%.[2] These numbers suggest that female students are at a disadvantage in obtaining permanent physics jobs despite their showing an interest in physics. Nevertheless, 10% is still higher than the world average.

We also found a significant difference between genders regarding marriage. Our survey shows that 5 of 24 female physicists are unmarried. These women range in age from 31 to 70 and have their careers either in teaching or research. For male physicists, only 4 out of 55 are unmarried in the same age groups. These numbers suggest that female physicists have difficulty balancing the family and career, partly due to social pressure, and partly due to individual constraints.

To promote the status of female physicists in Taiwan, we need to increase the number of female students in physics and related disciplines and create an environment that will encourage female physics graduates to stay in the field. We sincerely wish that our society could recognize this problem. One practical way to help solve this problem is to provide more supporting benefits and facilities. Public day-care centers for children and elders would greatly reduce the family duties for women. However, nothing is more important than raising public awareness about women's status in Taiwan.

REFERENCES

1. E Mayer, *China Times*, July 23, p. 37 (2001).
2. S. N. Yang, Physics Bimonthly (in Chinese) *21*, 605 (1999).

Colombia: The Place of Women Physicists in a Developing Country

Angela Stella Camacho

Universidad de los Andes

New and unexpected challenges must be faced and overcome in this millennium. Extraordinary advances in science and technology, especially in physics, during the 20th century have sparked considerable social changes worldwide, sometimes spurring progress and sometimes suffering for large groups of people. The effects of globalization are felt differently in different regions and countries. However, in most countries, globalization should improve the existing gender inequities in educational systems.

Colombia has only 28 years of scientific history. In developing countries such as Colombia it is extremely important to promote science and technology. We need to close the technological gap between us and developed countries. In this sense, science education acquires unprecedented importance, when we consider that present generations must be prepared to use science to promote sustainable development and improve the quality of life for millions of people. Efforts are being made to improve the interest of Colombian youth in science.

Women have an important role to play in these efforts. But as many studies since 1970 have demonstrated, women's presence in most fields of science remains unsatisfactory. Sexual stereotypes and conventional sexual roles continue in the educational systems of many countries. This is one of the more serious obstacles to motivating women to follow scientific careers.

Many governments have accepted and followed the recommendations concerning women scientists adopted at international conferences sponsored by the United Nations; however, much must still be done to increase the numbers of women scientists. In Colombia the total number of scientists (men and women) is still very low and the relative numbers of women to men on the different levels of the academic ladder are also very poor.

Women already play an important role in education, although as science educators we are still underrepresented. Our job extends beyond imparting scientific information. To be successful we must have a clear understanding of the gender bias of education and must be creative and resourceful in order to develop our own methods of conveying information. We must stimulate interest in the subject while enlightening our female students that it is perfectly possible to desire and pursue a career in science. The role of the female scientist is, therefore, to skillfully and purposefully impart scientific knowledge, while influencing the lives of her students in ways that will shift the existing paradigms and create openings for the greater and sustained involvement of women in science.

CP628, *Women in Physics: The IUPAP International Conference on Women in Physics,* edited by B. K. Hartline and D. Li
© 2002 American Institute of Physics 0-7354-0074-1/02/$19.00

The Past and Present of Physics in Croatia: Gender Differences in Graduation Statistics and Textbook Illustrations

Vjera Lopac, Andjelka Tonejc, and Planinka Pećina

University of Zagreb

The University of Zagreb has a long and honorable tradition, beginning in 1669 with the Zagreb Jesuit Academy, a three-year high-school where one could attend logic, physics, and metaphysics classes. The renewed University of Zagreb started in 1874. In 1876 the natural sciences were added to the university curriculum. The first woman attended physics courses in 1900, one year before women were officially allowed to enroll.

The Department of Mathematics and Science was separated from the Faculty of Philosophy in 1946, when the Faculty of Science was established. Today it has seven departments (Biology, Chemistry, Geography, Geology, Geophysics, Mathematics, and Physics). Smaller physics departments also exist at about 10 other (technical and biomedical) Faculties of the University of Zagreb, and within the other three Universities in Osijek, Rijcka, and Split. The Faculty of Science in Zagreb offers students a physics engineering degree, as well as high school teaching degrees in physics, and physics combined with computing, mathematics, or chemistry.

GRADUATION STATISTICS

During the 50 years between 1946 and 1995, 831 men and 406 women graduated in physics and geophysics. In the same period, M.Sc. degrees were obtained by 244 men and 74 women, and Ph.D. degrees by 160 men and 36 women. In Figure 1 we compare the numbers of male and female total physics graduates and Ph.D. physics graduates between 1986 and 1995. We could also compare the female student populations at different faculties over the years. In 1993/1994 at the Department of Physics, for example, 32% of the 111 students enrolled in the first year were female. For comparison, in 2001 the Faculty of Medicine enrolled 230 students, 64% of them women; five years ago 48% were women. In the same period, 20% of students enrolled by the Faculty of Electrical Engineering and Computing were women, whereas 15 years ago only 5% were women. Presently among the physics teaching staff, the percentage of women physicists is 17%. For the remaining teaching staff (young researchers, assistants, etc.) the ratio is 13%. By comparison, the percentage of women on the Department of Biology teaching staff is 50%.

Physics Graduates   **Ph.D. in (geo)physics**

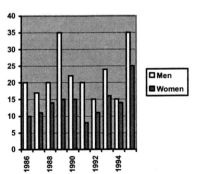
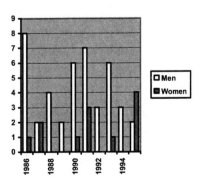

Figure 1

CP628, *Women in Physics: The IUPAP International Conference on Women in Physics,* edited by B. K. Hartline and D. Li
© 2002 American Institute of Physics 0-7354-0074-1/02/$19.00

How to continue to exist as a woman in physics? Most successful women physicists in Croatia are married (in about equal proportions to physicists and nonphysicists) and have children. Unlike the experience reported in some other countries[1], marriage to a fellow physicist seems to be a favorable, because living in the same environment implies better mutual understanding. For a woman working in physics, the support of family is essential.

TEXTBOOK ILLUSTRATIONS

Physicists of Croatia are greatly interested in physics education research. Many of them actively contribute to the education literature and write textbooks. Here we give results of an analysis suggesting improvements in textbooks to attract more girls and generally more pupils to the study of physics. Illustrations in the physics textbook make the content livelier and more interesting, but also offer to the student an opportunity for identification. For six general physics textbooks, of which four are in English (two algebra based[2,3] and two calculus based[4,5]) and two in Croatian (one algebra based for professional high schools[6] and one for elementary schools[7]), we investigated the following points: How many pictures in the textbooks contain human figures? Can the persons be recognized as masculine or feminine? What is the relation between the numbers containing feminine, masculine, and indistinguishable figures? Do these numbers vary by the subject treated in the textbook or by chapter? Are portraits of physicists included?

Some of the quantitative results are shown in the two diagrams of Figure 2. In most illustrations the pictures show people in some activity (sport, work, experiment) or observing some physical phenomenon. The student is encouraged to identify with the person shown, although it is sometimes difficult to discern the gender, which makes identification difficult. Sometimes the human figure is drawn as a kind of shadow, with no indication of the gender or any other personal characteristics. No identification here is possible—the human element is deliberately reduced. Offensive irony or deliberate caricature of female persons is sometimes present in the illustrations. The number of portraits of physicists varies greatly, but the women are extremely rare.

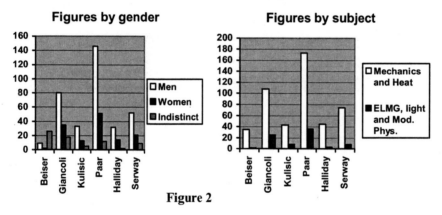

Figure 2

Our survey revealed that very few illustrations showing electricity and magnetism contain human figures. Instead, unattractive schematic diagrams prevail. It has been suggested that girls show less ability than boys in the field of electricity[8]; however, our investigation shows that nothing has ever been done to make these topics attractive for girls. To them the subject is totally new, because in most families the solving of practical problems with electricity is left to fathers and sons. Textbooks should depict both men and women in everyday situations, and parents should be educated about the need for girls to be involved in technical activities.

REFERENCES

1. L. McNeil and M. Sher, Physics Today (July 1999).
2. A. Beiser, *Modern Technical Physics* (The Benjamin/Cummings Publishing Company, 1978).
3. D.C. Giancoli, *Physics—Principles With Applications* (Prentice Hall, 1998).
4. R. Resnick, D. Halliday, and K.S. Krane, *Physics 1,2* (John Wiley & Sons, 1993).
5. R.A. Serway and J.S. Faughn, *College Physics* (Saunders College Publishers, 1999).
6. Ž. Jakopović, P. Kulišić, and V. Lopac, *Fizika 1,2,3,4* (Školska knjiga, Zagreb, 1995).
7. V. Paar, *Fizika 7,8*, textbook and exercise book (Školska knjiga, Zagreb, 1999).
8. T.R. Brown, T.F. Slater, and J.P. Adams, Phys. Teacher *36*, 526 (1998).

Women Doing Hard Sciences in the Caribbean

Lilliam Alvarez Díaz, Aurora Pérez Martínez, and Margarita Cobas

Institute of Cybernetics, Mathematics and Physics, Cuba

The purpose of this contribution is to present our perceptions of the common problems and main differences in the experiences of women working in the hard sciences in the various countries of the Caribbean Region. We define the Caribbean Region as all countries with coasts on the Caribbean Sea.

Up to now, sparse data and some analyses on the women-in-science issue have been published for individual countries in the Caribbean. Some international organizations, such as UNESCO, UNIFEM, and TWOWS, have developed and published data and statistics about women in science and focused on gender issues in general.

As women physicists and mathematicians in the Caribbean, we have analyzed which gender obstacles we are facing in our universities and research institutes. We also studied Cuba, Trinidad, Tobago, Guyana, and Jamaica to establish some basis of comparison. Cuba was studied in more depth—much more data are available for Cuba than for other countries—and this information was used as the basis for our analysis.

We identified problems common among Caribbean countries, listed below, as well as distinctions among countries with different socioeconomic systems and levels of social development.

The problems shared by Caribbean women in science include:

- Lack of leadership
- Tendency of girls to study traditionally female science subjects, such as the medical sciences and education
- Underrepresentation of women in the basic sciences such as mathematics and physics
- Migration and brain-drain (common for both males and females in the Third World)
- Underrepresentation of women at scientific decision-making levels
- Gender bias in textbooks, in the scientific vocabulary, and in the scientific literature (the "masculinity of science" is an issue even in the First World.)
- Stereotypes (Males are good scientists in physics or mathematics; males make good department heads; males make good deans or good chiefs!)
- Poor education in basic sciences at the primary and secondary levels

A wide range of socioeconomic conditions exists in our small countries of the Caribbean. Obviously, women doing physics and mathematics in Mexico or Brazil do not face the same situations as those in Guatemala or Nicaragua, where the social conditions are completely different.

We must also consider cultural traditions and social perceptions of the hard sciences in Third World countries, including the differences and similarities in traditions and perceptions among our countries.

We have concluded the following from our studies:

- The participation of women in all social and job levels, particularly in such fields as sciences, law, medicine, computer science, and engineering, has enormous implications not only for women but for the whole of our society and for the future of our Caribbean countries.
- By failing to encourage women, as well as men, to study the hard sciences, the Caribbean Region wastes a substantial portion of its talent pool and severely limits the potential contribution of women in these disciplines.
- Not only are Caribbean women underrepresented in the sciences, but their contributions are undervalued or perceived as marginal.
- The influence of role models of women doing physics and math is crucial; we must work hard to create an environment in which these role models can exist.
- The active support of policymakers and scientists is crucial to ensuring the full participation of women and girls in all aspects of the hard sciences and technology, and to overcoming existing gender stereotypes.

CP628, *Women in Physics: The IUPAP International Conference on Women in Physics,* edited by B. K. Hartline and D. Li
© 2002 American Institute of Physics 0-7354-0074-1/02/$19.00

Our recommendations are to:

- Make the publishing and distribution of biographies of Caribbean women in science a first-level priority in the Caribbean. We must present this material in our colleges and universities.
- Establish women's issues as a legitimate area of concern in higher education.
- Collect, analyze, and report gender-separated data, documenting constraints and progress in expanding the role of women in science in the Caribbean.
- Adopt a nonsexist, all-inclusive language policy to cover all written and verbal communications. Screen and edit curriculum materials for gender bias.
- Identify and fight the subtle forms of discrimination of women in sciences that are harder to prove and much harder to deal with and remedy.
- Realize that activism is the only way to change current situation and to correct the future.

ACKNOWLEDGMENTS

We are grateful to Prof. Grace Sirju-Charran and Veronica Broomes for their useful comments and information during our research.

How to Increase the Number of Women in Physics in the Czech Republic?

Raji Heyrovska,[1] Jarmila Kodymova,[1] Jana Musilova,[2] and Olga Krupkova[3]

[1]Academy of Sciences of the Czech Republic, Prague; [2]Masaryk University, Brno;
[3] Silesian University, Opava

Why is it important to increase the participation of women in physics?

The current participation of women in physics, especially in basic and applied research, is very low, reflecting the general situation in the natural sciences in this country. For example, the Institute of Physics of the Academy of Sciences in Prague employs about 220 scientific workers, only about 10% of whom are women. About 600 out of 3,000 students in the Mathematical-Physical Faculty of Charles University in Prague are female, and most of them are preparing for teaching careers. A higher number of women in physics and their participation in responsible academic positions, as in all of science, would benefit the scientific community because the female style of thinking, logic, diligence, feeling for organization of work, and systematic approach are different from those of men. In this country, access to education and work in the physical sciences is free for women, as it is for men, and increasing the number of women in physics will benefit our society's demand for physical and technical sciences and should be encouraged by the scientific institutions.

What are some of the joys and satisfactions experienced in a career in physics?

In sciences such as physics, the joys and satisfactions follow as much from enthusiasm and interest in the subject as from the fact that the work simultaneously is a hobby. This is true for both men and women. Satisfaction and appreciation of our female colleagues' work in the male-dominated field of has come from leadership of a research laboratory and its success in international collaborations, obtaining financial support from abroad, and invitations to deliver talks at significant international conferences and to be members of the advisory or program committees of international conferences.

Are there particular aspects of the culture in our country that made these possible?

Historically, the culture in the Czech Republic gives a good foundation for education, free opportunity, and professional careers for women. However, the low number of women working as physicists can be attributed to the demands of this profession while maintaining a family and caring for children.

Are there some exciting discoveries by our women physicists?

In view of the fact that financial support for research in the natural sciences in this country has not by far reached European standards, one cannot expect, especially in experimental physics, exciting discoveries. The number of women working in theoretical physics is rather low. Exciting discoveries of individual women scientists working in experimental physics is hardly possible at the current level of development of physics, research requires mostly teamwork. There are some scientific teams where women participate in the research work in various fields of physics, and these teams attain outstanding results on the international level.

CP628, *Women in Physics: The IUPAP International Conference on Women in Physics,* edited by B. K. Hartline and D. Li
© 2002 American Institute of Physics 0-7354-0074-1/02/$19.00

Are there cultural, societal, and/or professional features of a woman's career in physics that may be peculiar to our country or to a larger geographical area?

To be engaged full time in the physical sciences and be simultaneously a mother rearing children, a woman has to exert much more effort to catch up from the time spent on maternity leave (in this country, it can be 2–3 years), time off for children's illnesses, etc. For example, in the Academy of Sciences, the commissions do not take into account the above-mentioned women scientists' duties when evaluating performance or salary categories. The salaries here are so low that it is impossible to pay for outside help with household chores or child care. Also, due to other cultural and societal aspects, having outside help is quite unusual. All of these factors hinder women from academic work and some even give up this profession.

What have individuals or groups in our country done to try to improve the working conditions for women in physics?

In the context of the present unsatisfactory economic and social situation in this country, any improvement in working conditions, especially for women in physics, seems difficult. However, the seminar "Women in Science," held in Prague in May 2001, focused on the problems and perspectives of Czech women in science.

What are the most serious barriers to women achieving successful careers in physics?

Some of these are have been mentioned above. In addition, an inadequate appreciation of this profession in comparison with other more attractive and less demanding professions probably also plays an important role.

Are there ways in which physicists, both male and female, from our country and other countries might help to alleviate these barriers?

The alleviation of these barriers depends to a large extent on political and economic developments. Scientists in responsible positions have to convince politicians of the importance and necessity of giving equal opportunity for women. It is necessary to specially support and encourage capable women scientists and to help accommodate the family duties and catch up with the time spent due to family and child care. Due consideration must be given to women scientists who can devote more time for academic work after the family does not demand their time.

What initiatives have been taken in our country to attract more women to physics?

The field of physics in the Czech Republic needs more men and women, both. To attract more students to the study of physical sciences, the Academy of Sciences organizes "Open Door Days," when all research laboratories are open to high school students and to the public at large. The results of these efforts can be seen only in the future. No practical steps are taken to attract women in particular, although the recent Government Resolution No. 16, passed in January 2000, calls on the bodies responsible for the implementation of the science policy to *gradually* develop equal opportunity policies in all state-supported programs and in all grant agencies distributing public funds in R&D.

What does our team hope to contribute to and take away from this conference?

We wish to exchange information and experiences with women from different countries and cultural backgrounds. This should help us find more constructive answers to the above questions, which we can try to implement in our country.

Women in Physics in Denmark

Anja C. Andersen

Niels Bohr Institute for Astronomy, Physics and Geophysics, Copenhagen University

Measured on an international scale, Denmark has a high rate of women with careers. The existence of inexpensive and well-functioning day care centers has contributed to make this possible. Therefore, it is all the more puzzling that women are still not represented in significant numbers in some fields such as research, politics, and management.

In Denmark the first female full professor in physics was appointed in 2001. Among the permanently employed research physicists, fewer than 2% are women. In the pipeline of physics students, about 20% are women. This number has been constant for almost 15 years; before that the number was about half.

The lack of women in research positions is a general problem at Danish universities. There are more female full professors and permanently employed researchers in the humanities, but the recruitment pool is also significantly larger (often above 70%), so the problem of a lack of women is not on the same scale as seen in physics. The low representation of women in physics research in Denmark is part of a general trend of a low percentage of women in research. This causes a lack of role models, supervisors, and mentors for students and young female researchers. In order to increase female recruiting to research positions, it is necessary that more women show that it is possible to be a female researcher. It is therefore of crucial importance that the female element at all levels of the research hierarchy is increased, so that role models really are roles as well as models.

The fact that the number of women at the upper levels of the scientific hierarchy does not reflect the number of women educated in a scientific field implies that much potentially promising talent is lost from the universities. The female physicist who leaves the university career pipeline gets a job easily as a physicist (there is no unemployment in physics in Denmark), typically as high-school or college teacher or in private industry. The lack of women researchers in physics is therefore more of a problem for the universities than for individual women.

It is about time that the universities acknowledge that in order to recruit the most talented people in the workforce it is necessary to include women. Inertia, however, seems to be as evident at universities and relevant ministries as in Newton's laws of motion, so even today many universities frequently act as if they can afford to give up the talents of half of its citizens.

Affirmative action has not been practiced in Denmark to any significant degree, and it is not likely that it ever will be because there is a strong opposition among both men and women. It is considered undemocratic. The initiatives that have been made so far to alter the lack of women in university research careers have depended on a few top-level persons dedicated to the problem. For instance, in 1997 the (female) research minister made the lack of women in research an issue[1] and for about eight months it was debated in the media and at the universities. But due to a rearrangement of the government, the research minister was replaced after only one year and the next minister had other things on his agenda. So the results were only sporadic initiatives that faded quickly. A limited number of two-year "Curie scholarships" granted by the Natural Sciences Faculty of Copenhagen University, which made it possible for female researchers to renew their research qualifications, became victims of budget cuts.

Following the example of Norway, the Danish Physical Society has, in an attempt to promote women physicists, established a network for women in physics in Denmark (http://www.nbi.dk/kif). The main objectives are to:

- Create a network for exchange of information and knowledge among women physicists.
- Increase the visibility of women working in various fields of physics in Denmark.
- Increase the number of women studying physics and working in physics research by identifying gender barriers in the career paths of women in physics and by working toward removing such barriers.

---

[1] A report, "Women and Excellence in Research," can be found at the homepage of the Danish Ministry of Research, http://www.fsk.dk/fsk/publ/women/index.html.

- Exchange information on teaching methods that can enhance girls' and young women's interest in and benefits from physics and give women more self-confidence in their field.

These objectives are met through a newsletter (in Danish), an e-mail list, and meetings, joint colloquia, and seminars. The seminars and joint colloquia are open to the public.

In Denmark a great challenge for research policy in the coming years is that a large number of researchers will retire at the same time (about 2007). At this occasion, the universities are facing an unprecedented chance to change the fraction of women in physics research. If they do not take this opportunity, the prospects for changing the gender situation in physics in Denmark are bad.

Courage and Hope: Women and Physics in Estonia

Helle Kaasik

University of Tartu

More than half of the 1.5 million people of Estonia are women. The situation of women in physics is similar to that in many countries: women are remarkably underrepresented. The disproportion increases up the academic and administrative ladder, starting from very first steps of the education system.

The reasons for women avoiding or leaving physics belong to the realms of history, economics, and sociology. Specialists of these disciplines should be involved in studying the reasons and finding effective strategies for improving the situation of women in physics.

The history of gender-related issues in Estonia is multilayered. As the mostly foreign power over this country changed often, so also changed the position of women in society. Almost any distribution of tasks and rights between men and women can be claimed to be "the genuine Estonian tradition."

The old peasant tradition of Estonia considers women and men as different, but as equal partners. This tradition honored strength and intellect in both men and women. Women in Estonia were never possessions or objects of trade for men. Although a clear distinction was made between the tasks of men and women, there has always been a large area that both shared.

Later, the German-influenced urban family tradition of a breadwinning husband and caretaking, financially dependent wife classified women as secondary people. Several thinkers and writers in Estonia noticed this as injustice. The underlying tradition of equality and the remarkable general openness of our society to external influences made it easier for society to accept changes that gave women access to university education, voting rights, etc., leading toward the formally democratic society with equal legal rights for men and women that we now have.

When Estonia was a part of the Soviet Union, all men and women were equally obliged to work for salary, but child care and housework still remained mainly on shoulders of women—women got "the double load." Physics was strategically important (everybody remembered the nuclear bomb) and received good funding. To be a physicist was among the professions of highest prestige. Because the Soviet Union had free university education, an advanced social security system, and no unemployment, it was affordable and socially safe to become a physicist. Still, the percentage of women among the Estonian physics students remained low. As an interesting fact, the percentage was steadily and substantially larger in Russian-speaking departments of Estonian universities. (Some behavioral observations give the impression that the Russian subpopulation, reaching at its maximum about 40% of the total population of Estonia, generally had less prejudice toward female physicists than the rest of the society.)

With the decay of the Soviet Union, Estonia, after a short period of celebrating national freedom and independence, found herself in the early stages of capitalism. The importance of money grew, social security disappeared, both limousines and beggars appeared in the streets. Increased possibilities for achievement were accompanied by increased risk.

Physics lost its military importance, and obtaining funding for physics research became harder, especially for theoretical physics, which cannot yield immediate profits. All procedures for getting funding for physics changed. The meaning of "political correctness" also changed.

"The double load" for women continues, as legal pressure to work full time has been replaced by financial need. On the other hand, legal pressure on men to financially support their children has substantially diminished. A proposed new version of family law (not in force yet), despite its politically correct wording, brings harm to most married women. It takes from the poor and gives to the rich—as always, the winner makes the rules.

It takes courage to decide to become a physicist under such conditions. Being 18 and seeing the brave new world of capitalism, one has to have some special reason to decide to remain a poor student for next 4 + 2 + 4 years until Ph.D.— just to get the small salary of a junior fellow and start paying the loans he/she lived on. After long years of study it is not always possible to find a job in physics, not to mention a career. If everything goes by the straight line, our young fellow is 28 by that time and children still do not fit into the picture.

CP628, *Women in Physics: The IUPAP International Conference on Women in Physics,* edited by B. K. Hartline and D. Li

On the contrary, recognized senior physicists earn well and have great freedom in organizing their work schedule—good conditions for having children parallel with career. Here comes the unavoidable difference between men and women in science: for purely physiological reasons, it is much easier for men than for women to put off reproduction until their later years of life. Having children demands much more from mother than from father. So the widespread attitude expressed as "if a woman wants to become scientist, she should remain single" is based on bitter experiences of women.

Female physicists face a wide spectrum of societal attitudes ranging from admiration ("you must have an extraordinary IQ") to offending jokes. Behaviorally denying their sexuality and limiting themselves to the company of other scientists (who usually do not share these attitudes) are often defense mechanisms used by female physicists.

Physics is what physicists do, and if we lose half of the physicists we lose half of the physics. Fortunately some people, both men and women, still have enough courage and hope to choose physics for its beauty, which outweighs all obstacles.

Female Physicists in Finland

A. Penttilä[1], H. Aksela[1], U. Lähteenmäki[2], and J. Koponen[3]

[1]University of Oulu, [2]Centre for Metrology and Accreditation, [3]University of Helsinki

THE SITUATION IS GETTING BETTER

The number of female physicists is very low, but slowly increasing in Finland. As Table 1 shows, about 30% of the first-level degrees (M.Sc.) awarded in physics in the last three years were to women. This is much more than in the 70s, when only 10% of those graduating were women.

Table 1 also shows that the fraction of women continuing studies up to the Ph.D. level is lower than that of men. The most important reason for this decrease might be the setting up of a family: a teacher's profession is thought to be easier and safer economically than a career as a scientist. After all, getting married and having a baby is a greater change for a woman than for a man.

Another reason a woman may end her career as a scientist is to follow her husband to a new town because of his job, leaving her own job behind.

Other reasons are the difficulties faced as a member of the minority. Maternity leave cannot be taken for granted and a woman's family plans are always asked in a job interview. Despite the progress in the field of equality, prejudices still exist—more often among nonphysicist female coworkers. Nonphysicist women often view physics as a very difficult and boring subject and female physicists are sometimes treated as freaks. It is not a pleasure to work with people who consider you incompetent or strange.

The fraction of women in other fields of natural sciences is much better than in physics. In the fields of biology or chemistry men are the minority, whereas in the fields of mathematics or geography the number of women is equal to that of men.

TABLE 1. Degrees Awarded in the Universities of Helsinki, Oulu, Turku, and Jyväskylä.

Helsinki	2001 Women	2001 All	2001 %	2000 Women	2000 All	2000 %	1999 Women	1999 All	1999 %
M.Sc.	17	49	35	16	48	33	9	41	22
Ph.L.	6	18	33	2	13	15	2	13	15
Ph.D.	4	15	27	3	14	21	4	15	27
Oulu									
M.Sc.	9	20	45	10	21	48	5	28	18
Ph.L.	2	5	40	3	5	60	0	4	0
Ph.D.	1	4	25	0	8	0	1	7	14
Turku									
M.Sc.	5	29	17	5	23	22	3	26	12
Ph.D.[a]	1			0		0	0		0
Jyväskylä									
M.Sc.	3	42	7	6	36	17	5	37	14
Ph.L.	0	2	0	0	2	0	0	1	0
Ph.D.	1	11	9	1	9	11	1	5	20

[a] We were not able to obtain exact numbers from Turku. Annually there are fewer than 10 dissertations and the number of women receiving them is between 0 and 2.

CP628, *Women in Physics: The IUPAP International Conference on Women in Physics,* edited by B. K. Hartline and D. Li

TABLE 2. Positions at the Universities of Helsinki, Oulu, Turku, and Jyväskylä.

Helsinki	Women	All	%
Professors	0	23	0
Senior Assistants	1	12	8
Assistants	3	15	20
Oulu			
Professors	1	10	10
Senior Assistants	1	11	9
Assistants	3	14	21
Turku			
Professors	0	13	0
Senior Assistants	0	5	0
Assistants	2	12	17
Jyväskylä			
Professors	1	15	7
Senior Assistants	0	5	0
Assistants	1	6	17

PROBLEMS ORIGINATE FROM SCHOOL

The reasons for such a low fraction of women studying physics can be looked for in the upper secondary schools. Pupils' opinions on physics have been studied[1] and girls were found to consider physics an obstacle more often than boys. Also, the examples in the course books are more familiar to boys than to girls.

In the study physicists were found to be difficult-to-understand, theoretical, lonely workers, advanced in years and enthusiastic about their field. Physics itself was found to be mathematical, diagrammatic, laborious, hard to understand, and experimental. Concepts like international, modern, logical, or practical were not really connected to physics.

These ideas might follow from a lack of knowledge—there are hardly any physicists known to the general public, even less so for female physicists— as can be seen in Table 2. Almost the only example of a physicist for pupils is the physics teacher in the school. The study found that the first physics teacher was found to have the most important effect on the desire to study (or not to study) physics in the future.

The differences between universities can also been seen in Table 2. The universities founded more than 100 years ago (Helsinki, Turku) have more men in positions than the younger universities (Oulu, Jyväskylä). It is also worth mentioning that most of the professorships are occupied by soon-to-be-retired men. The change of generation might be an opportunity for female physicists in Finland in a few years.

REFERENCE

1. J. Häkkilä, M. Kärkäs, H. Aksela, V. Sunnari, and T. Kylli, "Girls, boys and physics. Opinions of upper secondary school pupils on physics as a subject," Series A, No. 14 (Publications of the Educational Office, University of Oulu, Oulun yliopistopaino, 1998).

The Status of Women in Physics in France

Etienne Guyon[1], Claudine Hermann[2], Martine Lumbreras[3],
Anne Renault[4], and Monique Schwob[5]

[1]*École Normale Supérieure;* [2]*École Polytechnique;* [3]*Université de Metz;* [4]*Université de Rennes 1;*
[5]*Institut National de Recherche Pédagogique*

The French delegates to the IUPAP International Conference on Women in Physics proudly acknowledge the choice of Paris for this meeting. France has a long tradition of promoting women in science, particularly through the elite education of women in the Ecoles Normales Supérieures de Jeunes Filles (see below).

THE PRESENT SITUATION

In secondary education the number of girls is about equal to boys: 55% girls are in nonspecialized *2ème* classes (third year before the end of *lycée*) and 43% are in scientific *terminale*, the last year in secondary education. When choosing "hard" sciences in higher education, the percentage of women drops significantly (see below). The origin of such a situation seems mainly cultural.

France has a dual higher education system: (a) the universities, in which the percentage of women students in 1998 at first-level degree was 37% in material sciences and 21% in engineering; and (b) engineering schools and ecoles normales supérieures (ENS, where future secondary and university teachers or researchers are educated). Admission into such schools is based on competitive entrance examinations, prepared for in two or three years of classes: in the classes focused on mathematics, physics, and chemistry, women comprise about one-fourth of the students. The proportion of women reaches one-half for the life sciences. Approximately 25% of engineering diplomas are now awarded to women.

The percentage of French women physicists in academia, amounting to 26% for university maîtres de conférences (assistant professors) and only 8.8% among professors, places France in a reasonable position among European states. In physics, depending on the subdiscipline, women's participation can be very low (particle physics and mechanics) or much higher (atomic or condensed-matter physics).

PECULIARITIES OF THE FRENCH EXPERIENCE

The present status of French women physicists is not determined by affirmative action or quotas; these are in general not appreciated, the fear being that a woman put in a position for quota reasons would be devaluated. Instead, the present status reflects the general French social situation.

General Social Situation

Women's wages in France are on average 25% lower than those of men; moreover, women only represent 35% of executives. Despite these facts, the social situation for working women is rather favorable, and it is considered normal for a mother of young children to maintain professional activities. An infant-care system exists; école maternelle (public preschool system for children from 3 to 6 years of age) is free of cost, takes place during the mornings and afternoons (except during holidays), and is attended by more than 90% of all children, even at 2 1/2 years of age. In addition, the law permits parental leave from work, an option mostly used by mothers, and tax deductions are applicable for child-care costs (until the child is 6 years old). These advantages, although not ideal, do not exist in all European countries.

CP628, *Women in Physics: The IUPAP International Conference on Women in Physics,* edited by B. K. Hartline and D. Li
© 2002 American Institute of Physics 0-7354-0074-1/02/$19.00

Universities and Research Institutions

The scientists in French universities or research institutions are in general under civil servant or state-employee status. They are hired for a permanent position between 30 and 35 years of age, depending on the discipline, after a short postdoctoral period.

Generally speaking, geographical mobility is very low in France for all professions. The dual-career problem is a common issue for many teachers and researchers: there is no definite obstacle, but it is not particularly organized. Although mobility is theoretically praised in academia, in reality it is only considered positively and recommended when one is being promoted to professor.

There is no discrimination in the salary structure between men and women, the fact that women receive fewer promotions explains the lower average wages quoted above.

An Example of Coeducational Difficulties

ENS de Jeunes Filles, founded in 1881 to educate future women teachers for the girls' public grammar schools, became coeducational in 1986 by merging into ENS de la rue d'Ulm. The alumnae of ENS de Jeunes Filles who studied there in the mid-60s are now mostly university professors or research directors at the Centre National de la Recherche Scientifique (CNRS), with a very reasonable success compared with their male colleagues. At that time, the selective entrance examinations for men and women for ENS were distinct. The initial separation of boys and girls was a way to protect the fairness of the percentages of male and female students (60% and 40%, respectively).

In the 1980s the decision was made to combine the entrance examinations for men and women. This had a dramatic effect. The number of girls who passed the entrance examination immediately dropped in mathematics and physics: the percentage of female students is now under 15%—a little smaller than the proportion of girls in the classes preparing for the very selective entrance examination. Indeed, a generation of women researchers has been lost!

WOMEN'S PRESENCE IN PHYSICS

The most famous French woman physicist was Marie Curie, born in Poland, who received her Ph.D. and had her career in France. The French Academy of Sciences now has 5 women out of 139 members: one of them is a physicist, Marie-Anne Bouchiat, at the origin of the verification of parity violation in atomic physics.

The barriers faced by women are mainly in getting promotions and in representation on decision-making committees. There has been strong recent political support from the ministries of education and research, and women have been nominated as directors of research institutions. For example, the head of CNRS is now a woman, as well as the head of the Physical and Mathematical Sciences Department of CNRS. At the European level, Catherine Cesarsky, an astrophysicist who spent a large part of her career in France at the Atomic Energy Commissariat, is now the director of the European Southern Observatory.

As in other countries, it would be beneficial to increase women's participation in physics: this would lead to interesting jobs, both inside and outside of academia. Greater participation of women would bring complementary approaches to science or technique and contribute to a good climate in the work environment.

This question of the presence of women in physics has been recently highlighted in the context of the crisis of scientific vocations: physics is the most concerned sector in France, and more women would be heartily welcomed!

The Status of Women in Physics in Germany

Corinna Kausch[1], Barbara Sandow[2], Monika Bessenrodt-Weberpals[3], and Silke Bargstaedt-Franke[4]

[1]Gesellschaft für Schwerionenphysik; [2]Freie Universität; [3]MPI für Plasmaphysik; [4]Infineon Technologies AG

From 1993 to 2000, the proportion of women/men in physics at different levels of education in Germany was as shown in Table 1. The percentage of women teaching physics in universities is growing, but very slowly (Table 2).

TABLE 1. Percentages of Female Physics Students.[1]

Education Level	1993	1994	1995	1996	1997	1998	1999	2000
Undergraduate Studies: Beginning	13.8	16.1	17.4	17.9	20.0	21.2	22.5	22.5
Undergraduate Studies: End	9.1	9.1	9.2	9.1	10.5	8.8	9.1	9.8
Ph.D.	8.0	7.3	7.9	8.4	8.1	9.4	9.9	9.9
Habilitation (Advanced Postdoctoral Level)	6.7	2.7	5.0	3.1	3.6	3.6	7.8	7.7

TABLE 2. Percentages of Woman Teaching Physics in Universities.[1]

Staff Category	1992	1993	1994	1995	1996	1997	1998	1999	2000
Faculty Staff	6.7	6.9	6.8	7.7	7.4	8.0	8.5	9.4	9.7
Professors	1.3	1.3	1.2	1.4	1.5	1.5	1.6	1.7	2.8

The most common reasons for women in Germany choosing careers other than research after having studied physics are:

- Lack of permanent positions for young scientists
- Better career options and salary in other professions
- Domination of the academic world by an extremely male system
- Practical issues in combining family/research
- Lack of positions for two-career couples in physics at the same research institute

In scientific careers, a long period of probation is required before a permanent job is obtained. Many women disappear from research during the postdoctoral time when scientists get only short-term contracts. This is also the time to have a family, and it is difficult for a woman to manage both a family and a career.

The traditional role of a woman in Germany is that of being a mother, which poses a very severe barrier to achieving a successful career in physics. Society still thinks that a working mother with a small child is a bad mother, and that a father who reduces his working time to care for his child is unable to pursue a career. Since childcare is usually a private problem, a very supportive husband is necessary.

Physics is known as a profession that cannot be combined with a family. Girls to whom having a family is important do not choose physics!

CP628, *Women in Physics: The IUPAP International Conference on Women in Physics*, edited by B. K. Hartline and D. Li

GERMAN PHYSICAL SOCIETY SURVEY SHOWS DISCRIMINATION

In 1998 the Committee for Equal Opportunities (Arbeitskreis Chancengleichheit: AKC) of the German Physical Society was founded. The AKC is dedicated to achieving equality and full participation for women in physics. The activities of the AKC are manifold and focus on women in physics at all stages of their careers:

- The Hertha Sponer Prize for young female physicists
- An annual female physicists conference with more than 200 participants
- Sessions focusing on women in physics at the conferences of the German Physical Society
- Projects to attract girls into physics and workshops to coordinate university projects
- Associations with other organizations in chemistry, engineering, etc.
- Plans for special workshops in career training and combining children and career

The AKC initiated a survey in the year 2000 that was sent to all female (3062) and 800 male members of the German Physical Society to assess whether equal opportunities exist in physics working areas.[2] The survey found that women are discriminated against, especially in the governmental research institutes (e.g., universities). They are not promoted at the same rate as their male colleagues. Even when women show their talent during their studies, they remain under a cloud of suspicion because they may have to care for a family and require a reduced workload. This reduced workload seems to be expected from the women only, whereas their partners are usually physicists also (54.6%). Almost twice as many female physicists (87%) have partners with an academic education, compared with 47% of the men.

Seventy percent of female physicists do not have children, compared with 49% of male physicists. But even women who do not have children do not have equal opportunities. They remain in lower positions, get promoted less often, and earn less money. Women of all ages are found less often in high-level positions. At age 45 and above, only 44% of the women occupy high-level positions in physics, whereas 73% of men do.

Even when women are in leadership positions, their salaries are lower. In their early careers, women experience few disadvantages and these seem to be incidental. However, with increasing experience, women realize that they have fewer opportunities. The result is a growing discontent with their job situation, whereas men are more content with their working situation with increasing age.

Women who have children are almost as successful as women without children. They have comparable career opportunities. This discrimination is not conditional upon their decision to have children but only because they are women.

The political climate in Germany changed recently with the appointment of a female minister of science and education. This has led to many national political efforts, such as campaigns to win girls' interest in becoming engineers and physicists. These endeavors have only a small impact, because in Germany physics is a branch with high social merit; as a consequence, it is dominated by an extremely male system. This system, as well as factors such as late seminars, the expectation of a 60-hour work week, and short-term contracts, is responsible for the low participation of women in research. It is an enormous waste of excellence that there are many women educated in physics who, due to unfriendly working conditions, stop working in physics!

REFERENCES

1. Statistisches Bundesamt (Federal Statistical Office of Germany), Wiesbaden (2001).
2. B. Koenekamp, B. Krais, M. Erlemann, and C. Kausch, Physik Journal *1* (2), 22-27 (2002).

Women in Physics in Ghana: A Situational Analysis

Aba Andam

University of Ghana

PARTICIPATION OF WOMEN IN SCIENCE IN GHANA

The status of women in physics in Ghana is best considered in the context of the general position of the participation of girls and women in scientific studies and careers. In Ghana, as in many countries worldwide, science has been perceived in the past as a difficult subject which is best left to a select few, particularly men. As a result of intervention strategies put in place by the Ghana Education Service, awareness is being created within the society, and a new emphasis is being placed on science studies in general for girls and on physics studies specifically. Nevertheless, a lot more needs to be done to get women into physics careers in Ghana.

Women make up less than 20% of the scientific workforce in Ghana.[1] The majority of women scientists in Ghana are in the pure and applied biological sciences such as biology, biochemistry, agriculture, food science, pharmacy, and nursing and medicine. It is observed that the majority of women in science are in the lower echelons of responsibility and decision making. In the teaching profession, for example, women are most often lecturers, researchers, or assistant researchers, and seldom professors or research directors.

SOME REASONS FOR THE IMBALANCE

The main reason for the paucity of women in scientific careers in Ghana is that there are simply not enough girls studying science. This is a direct result of societal attitudes, reflected in the school, that studying science is essentially for boys, who, upon graduating, will be prepared for the world of work that is perceived to be masculine.

Science is not portrayed as a subject that is vital for daily living. Science is often portrayed as a subject that belongs to the classroom only, for which many years of study are required in order to achieve a qualification that will be used for a career outside the home. The general attitude toward girls is that their future roles as mothers, home managers, and perhaps members of the general workforce do not require long years of science study. Some studies in Ghana and other African countries have also revealed that the method of teaching science contributes to killing girls' interest in science. [2–4]

With these negative influences around them, many girls accept the status quo and internalize the intellectual boundaries set for them by society. Even some brilliant girls give up easily in the face of the opposition they meet from schoolmates, teachers, parents, and relatives when they try to study science.

WOMEN IN PHYSICS IN GHANA

Physics holds a place of pride in the sciences because it is central to understanding the sciences. However, in Ghana, physics has been studied by many brilliant students in secondary school merely as a subject to pass in order to gain entry into higher education to pursue applied science courses such as medicine and engineering. The industrial backbone of the country is not strong enough to absorb qualified career physicists.

There are very few women physicists in Ghana. The total number of women who have taken a first degree in physics is less than 100, in a population of nearly 18 million. The first Ghanaian woman to gain a university degree in physics graduated with B.Sc. in 1973. There were many intervening years in which no woman graduated in physics.

An increase in recent years has come about as the result of a general intervention program initiated by the Ghana Education Service.

CP628, *Women in Physics: The IUPAP International Conference on Women in Physics,* edited by B. K. Hartline and D. Li

Since 1987, the Ghana Education Service has worked to raise awareness about the participation of girls in science studies. Its Science Clinic for Girls has achieved positive results in making girls aware of the different options open to them for scientific studies and careers, including physics. The result has been an increase in enrollment of girls for physics programs in universities. The increase is modest, but it is hoped that it is the beginning of a new era of appreciation of physics as a worthy career for women.

REFERENCES

1. Ghana Statistical Services Data, 2001.
2. A.A.B. Andam, *Toward a gender-free science education in Ghana,* proceedings of Forum on Science in Sub-Saharan Africa (American Association for Advancement of Science, Washington D.C., May 1993).
3. L. Dimandja, *African Women—Their Training and Position in the Fields of Science and Technology,* proceedings of ICWES9, The Ninth International Conference of Women Engineers and Scientists (Warwick, UK, July 1991).
4. L.P. Makhubu, *The Potential Strength of African Women in Building Africa's Scientific and Technological Capacity,* Keynote Address at Forum on Science in Sub-Saharan Africa (American Association for Advancement of Science, Washington D.C., May 1993).

The Status of Women in Physics in Greece

Christine Kourkoumelis[1], Eleni Pavlidou[2], Chara Petridou[2], and Eleni Stavridou[3]

[1]University of Athens, [2]Aristotle University of Thessaloniki, [3]University of Thessaly

UNIVERSITY SITUATION

Greece is a country where the number of women in undergraduate physics is particularly high (36%). At the graduate level, this number drops to about 22%. A similar percentage—about 20%—holds true for women with academic careers in physics (see Figure 1), but it is *almost 0%* at the highest level (full professor). As a reference, the percentage of women in senior management in all fields in Greece is about 6%.

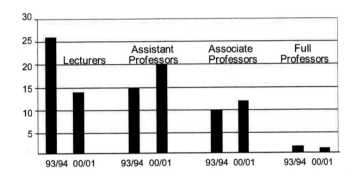

Figure 1. Percentages of women physicists in each academic rank for two different time intervals.

The numbers are too low: very rarely have women managed to reach the full professor level or directorship of a research institution. Furthermore, the money spent for research in Greece is the lowest in Europe, which makes things difficult even for men.

We have observed that most women graduate from the university with a bachelor's degree and afterward enter the job market as high-school teachers (mostly private schools or tutoring). Because men in the same age group have to serve for two years in the military, women clearly have a time advantage, but the advantage is later lost because of time off due to pregnancy, child bearing, or other family duties.

SOCIAL SITUATION

In Greece the social situation is such that both members of a couple have to work to support the family. Women physicists prefer teaching, which has a more flexible schedule. Building a career in research often means—as in most countries—delaying or giving up family life.

If a woman works in the public sector, legislation protects her in family issues, including a two-plus-three-month leave of absence because of pregnancy and a prohibition against firing her. On the other hand, she loses contact because of the reduced time available for work and traveling. Part-time employment and job-sharing, although legally established three years ago, are not available in the public sector. In the private sector things are less certain, especially because private tutoring is often done unofficially.

The usual barrier to successful careers for women in Greece is that men at the top ranks have tailored these jobs to fit in their life-style. Also, these men cannot often understand that women can be:

- Clever
- Able to handle flexible work schedules
- Reliable
- Bosses(!)

CP628, *Women in Physics: The IUPAP International Conference on Women in Physics,* edited by B. K. Hartline and D. Li

GROUPS INVOLVED IN IMPROVING THE SITUATION

According to the Hellenic Union of Physics, which consists mainly of physics teachers, no study on women and their careers in physics in Greece exists. A wider study, "Report on Trends of Female Employment in Technical Occupations in Germany, Greece, Finland, France and United Kingdom (1993–1999)," found that physics is considered an "occupation with high technical job content," and that overall there is a 1% annual increase in these occupations following an improvement in average educational level, irrespective of marital status.

Matters concerning gender equality are handled through the General Secretariat for Gender Equality (http://www.kethi.gr/english/), which is supervised by the Ministry of Interior, Public Administration and Decentralisation. The Secretariat is responsible for promoting and supporting legal and substantive equality in all sectors (political, economic, social, cultural). The Secretariat proposes measures to be taken by the State, collaborates with international organizations, promotes actions, intervenes with various political bodies, and issues updates and information. Its current priorities are harmonized with the practice and developments of the National Employment Plan; the planning of the Third Community Support Framework; and the new provisions on equality in the European Union Treaty, the European documents regarding the promotion of women in decision-making centers, and the course of the European policy on equality.

Specifically, the Secretariat's priorities are:

1. Gender mainstreaming
2. Strengthening the mechanisms promoting equality
3. Work/employment
4. Balanced participation in decision making
5. Ending violence against women
6. Mass media/publicity

PRIMARY AND SECONDARY EDUCATION

In Greece, elementary-school students are first introduced to physics during their 5th and 6th years. The majority of teachers at these levels are male. The female teachers prefer to teach to the first three grades. This is partially due to the fact that the early grades need less preparatory work, which leaves more time for housework/family. As a consequence, both girls and boys—our schools are coeducational—get the message that physics (and science) is a man's affair or that men are more capable of dealing with technical/technological/experimental matters.

Some members of our team have studied the problem and have suggested a number of interventions that could decrease the gender bias in elementary schools and provide equal opportunities to both girls and boys in science education.

On the other hand, statistics show that girls are doing equally well in these introductory-level courses. One study shows that during the 5th grade, 55% of boys and 45% of girls consider physics a "very interesting" subject.[2]

The data we were able to gather about education in high schools indicate that teaching is done by both male and female teachers, but that female teachers prefer teaching the lower grades of high school (high schools instead of lyceums). The physics teaching, though, is not done well. As a result, student interest in physics drops continuously during the high-school years. A year before graduation, 70% of the boys and 89%(!) of the girls consider physics as "not interesting at all."[2]

REFERENCES

1. "Report on Trends of Female Employment in Technical Occupations in Germany, Greece, Finland, France and United Kingdom (1993–1999)," http://www.kethi.gr/english/MEleti_Leonardo/periexomena.htm.
2. *Proceedings of the First Panhellenic Conference on Teaching of Sciences and Application of New Technologies in Education* (in Greek), Thessaloniki, Greece (1998).

Physics in Hungary

Judit Nemeth

L. Eotvos University

Physics is taught at four universities, two of them in Budapest and two in small countryside towns. There are two career possibilities for students studying physics in Hungary: teacher or researcher. The great majority of the students studying to be physics teachers are girls, while among the research students only 10–12% are women. The university training for research students and the standard of the lectures given to them are much higher.

Officially there is no effort made to encourage girls to become physicists. On the other hand, women realize quite easily that they have to overcome serious difficulties if they want to become researchers.

Hungary is a small country with very few research positions. In order to be able to compete for these positions with their male colleagues, women must go abroad for years for Ph.D. studies and for postdoctoral positions in order to gain experience in new fields of research. They can usually only do this if they are not married, or if they have a physicist husband working in the same field (the chance of both partners getting a position at the same institute is very small). Even when a woman succeeds through the postdoctoral level, to continue to be competitive, she usually cannot have a family before the age of 32–35.

If these conditions are not fulfilled, the woman physicist's hope for an interesting career is very small. Industries in Hungary are only now starting to employ physicists. The majority of women physicists who finished university training in recent years have either left physics and work in other fields, or work as experimental physicists—but as a team member, not a team leader. A minority decide to postpone marriage and children for a later time; the most successful women physicists are among these. Only a very few of the women physicists who try to have a normal family life as well as a career are successful.

Low salaries in countries like Hungary make everyday life extremely hard. However, we think that the obstacles we mention above for women in physics are not unique to Hungary but experienced by women physicists worldwide, and that it is very difficult to overcome them. One of the things our team would like to find out during this conference is whether the same problems arise in other countries, and if so, how women are able to overcome them.

For the above reasons, we do not believe it is important to increase the number of women in physics, because there are many fields in which women can succeed more easily. On the other hand, we believe society should help those who have a special inclination toward physics.

Because there are very few women in research, we cannot recall a single discovery made by women physicists working alone. There are some very good research pursuits in atomic physics, statistical physics, astrophysics, and nuclear physics, and in some of them the leading role belongs to women. We do not believe that this work is important for our country, but simply important for physics.

There is one area in physics where, at least in Hungary, women are very successful: science organization. We are very successful in organizing workshops, conferences, working teams, and high-school student groups.

We think that the joys in our careers are the same as for our male colleagues: to solve problems successfully, to have good students, and to organize good workshops. In our views of success, men and women are the same.

CP628, *Women in Physics: The IUPAP International Conference on Women in Physics,* edited by B. K. Hartline and D. Li
© 2002 American Institute of Physics 0-7354-0074-1/02/$19.00

Women in Physics: An Indian Perspective

Neelima Gupte[1], Jyoti Gyanchandani[2], Sunita Nair[3], and Sumathi Rao[4]

[1]Indian Institute of Technology, Madras; [2]Bhabha Atomic Research Centre; [3]Raman Research Institute; [4]Harish-Chandra Research Institute

Understanding nature, which is what physics is all about, is a global human endeavor. However, for a variety of reasons, women are conspicuously underrepresented in physics. In the new millennium, it is certainly important to correct this underrepresentation—both for the sake of the women, who should have the same opportunities and resources to realize and fulfill all they desire to accomplish in their lives, as well as for the sake of physics, which must draw on the full potential of humankind to solve its myriad questions.

Within the Indian context, the participation of women in public life has been quite high. Even before independence, thanks to an enlightened Indian leadership, especially that of Mahatma Gandhi, both men and women have been involved in the freedom struggle. This led to the emergence of strong women leaders in the subcontinent. Indian women have held most political positions of importance, such as prime minister, chief ministers of various states, and leaders of legislatures. Women have also held positions of prominence in the judiciary and in other professions.

However, India is a country of many contrasts. Along with women of very high levels of accomplishment, it also has problems such as high levels of female illiteracy, female infanticide, and dowry deaths. This leads to unusual statistics. For instance, although almost half the women in India are illiterate, roughly a third of science graduates are women and a reasonable fraction of them stay on in science.

It is interesting that a survey of all the working scientists in the country reveals that, unlike in the West, there is very little perception at either school or college level that women cannot do physics and math. The more standard reasons for dropping out seem to be family or marriage related. It becomes difficult for women to cope with the simultaneous time demands of their gender roles and their professional commitments. While training up to the Ph.D. level puts heavy demands on the time and commitment of women vis-à-vis the demands of society, which expects them to fulfill the commitments of marriage and raising a family during the crucial years of training, postgraduate degrees in science are common, and are even encouraged by parents. Moreover, jobs such as college teaching or working in scientific establishments after a Master's degree are perceived as highly desirable and prestigious for women.

However, in graduate school and beyond, the dropout percentage of women becomes far more significant. Whether or not a woman completes a Ph.D. and/or pursues postdoctoral fellowships is completely dependent on her marital status, the employment of the spouse, and family support. The women who go ahead for a Ph.D., postponing marriage, often marry other students. This leads to the problem of finding two jobs in the same place. So either the couple has to compromise on the career of one of them (usually that of the woman), or end up with many years of a commuting marriage. Moreover, most jobs in academics require many years of postdoctoral training. This again entails commuting marriages, postponing or coping with the additional problems of child-bearing, and dealing with a great deal of family pressure.

Even after all this effort, suitable academic jobs are difficult to get. Those who do manage to get suitable jobs along with their spouses still have to deal with problems of child-rearing, given the fact that suitable day-care centers and crèches with educated caregivers are still relatively new concepts in India.

In spite of these handicaps, women physicists in India form roughly 10% of the physics faculty in all of the universities, and have significant presence as researchers in the various government institutions and research laboratories. This is a pretty impressive achievement.

Additionally, many women physicists have been working on frontier problems at the international level. In particle physics, Indian women physicists were involved in the top quark discovery and theorists have written highly cited and useful papers in phenomenology and quantum field theory. In fact, one of the most highly cited particle

CP628, *Women in Physics: The IUPAP International Conference on Women in Physics*, edited by B. K. Hartline and D. Li

physicists from India is a woman. In astrophysics, they have authored well-received papers in galactic dynamics, plasma physics, and physical cosmology, and have been involved in the development of India's various observational facilities. Two women astrophysicists have been appointed directors of planetaria.

In condensed-matter physics and nonlinear physics, some of the most active researchers in the country are women, and they have contributed extensively to the fields of strongly correlated electron systems, nonlinear dynamics, and statistical mechanics. Many women have made solid contributions in the areas of nuclear physics, material science, and applied physics. However, it is still true that recognition of the achievements of women physicists is small. There have been extremely few prestigious awards or fellowships in academies of sciences awarded to women physicists in India.

One can now ask what can be done for the betterment of women's careers in physics. At the college and high-school levels, there are many students, both boys and girls, who are interested in science in general, and physics in particular, but are pushed toward professional courses by peer as well as parental pressures. To offset this, scientific discoveries and profiles of scientists need to be highlighted in the media.

We need to provide support for high-school and college teachers so they can expose their students to the possibilities and options of careers in basic as well as applied physics. The best way to do this may be to help teachers maintain contact with current research. This would both help the very many extremely competent women who teach at this level to maintain their own scientific levels and careers, and enable them to enthuse students toward careers in physics.

To attract young women into Ph.D. programs and beyond requires some special efforts. In particular, it needs a certain sensitizing of society to the view that both men and women should share responsibilities for home and child care. Unfortunately, although women have become more progressive over the years in India, men, and society in general, have not kept pace. So special efforts are needed to involve the media in debunking stereotyped roles for women and men, and in the general "sensitizing" of the male half of the population.

To help women through the crucial child-bearing and -rearing years, perhaps schemes of part-time work (with part-time salaries) can be instituted for both women and men, but with the option of being later converted to a normal full-time position. Another possibility is to allow reentry into the career path after taking time off, which would require that age restrictions for various positions be dropped. It should also be possible for women who perform well to have faster tracks after reentry.

A simple way to ensure that the special problems of women are looked at with empathy and insight is to appoint women members to evaluation committees. It is also necessary to have strong guidelines preventing institutions from undermining efforts toward affirmative action. Although the Indian government does have strong rules discouraging sexual harassment at the workplace, and many institutions have panel's to address gender-specific discrimination and sexual harassment, much more needs to be done toward making the work environment safe and women-friendly at all institutions.

Recognition, mentoring, and rewarding of women achievers is also required. This is needed not only for the morale of the women in a male-dominated field, but also because these women serve as role models for new generations of women scientists.

If men and women are the two pillars on which society is built, a fair, sensitive, and mature society will ensure that both these pillars are strong and receive justice. We hope the discussions at this conference will lead to further concrete efforts toward this end.

Physics for Women in Indonesia

Wiwik S. Subowo

Indonesian Institute of Science

When I received the first announcement about the IUPAP Women in Physics Conference, I felt very happy because a group of physicists is paying attention to the existence of women in physics. I felt happy corresponding with many women physicists from other countries, such as Judy Franz, Marcia Barbosa, Ling-An Wu, and Katie Stowe, and I am delighted to be able to attend the conference. However, I was surprised when I read the questions on the IUPAP Women in Physics survey, which will allow comparison between men and women physicists in any aspect of their careers (see survey results in these Proceedings). I hope the survey results will indicate that no difference in treatment exists between men and women in their careers as scientists, and especially as physicists.

Although the number of women physicists in my country is far less than the number of men physicists, and only a few women hold a Doctorate in Physics, I am sure that it is not because of different treatment of women physicists, but rather caused by personal constraints. It is a fact that women have more responsibilities in their household compared with men—especially in raising and educating the children—and this is natural. Only in special cases where the husband supports his wife in pursuing a doctoral program and sees the program as the family's program, will everything be well. Of course, she has to work hard to achieve the target. For unmarried women, it is easier to pursue a doctoral program, but generally they are afraid that they will lose their chance of getting married. I think that is the main constraint that keeps the women in my country from pursuing a doctorate degree.

PHYSICS EDUCATION

I am a researcher in the Research Center for Physics at the Indonesian Institute of Sciences. I have been the person in charge for the Indonesian Physical Society, Bandung Branch, from 1998 to 2000. Although I am not a teacher or lecturer, I am concerned with physics education, because many high school students complain that physics is the most difficult subject and is not interesting to them. So how do we overcome this problem? I formed a small team of Indonesian Physical Society members who were concerned with physics education in the high schools of Indonesia. The team consisted of high school teachers, lecturers, and researchers. They met every two weeks, and discussed solutions for this problem. We also pursued this problem at the 12th National Congress on Sciences held in Serpong, Indonesia, in September 1999.

We looked at the problem of physics education without differentiating between men and women or boys and girls, since we believe that there are generally no gender problems in education. Basically, by studying physics, the students will discover some remarkable things, namely:

- From universe, galaxies, and planets to the small systems in molecules, atoms, and nuclear particles, all are in order and follow the rules of physics. It makes us admire God the Great Creator. Hopefully, this feeling will not make people arrogant.
- The physics phenomenon can be found in all aspects of daily life: for example, physics in the household, physics in sports equipment, physics in vehicles, physics in engines, physics in medical tools, physics in toys, and so on.
- Basically, to learn physics is to exercise the brain, to think systematically, to develop analytical skills, and to increase creativity.

CP628, *Women in Physics: The IUPAP International Conference on Women in Physics,* edited by B. K. Hartline and D. Li
© 2002 American Institute of Physics 0-7354-0074-1/02/$19.00

If we want to prepare women to specialize in physics, I suggest the theme, "To increase the ability of women to think systematically and creatively through physics education." Women should be smart and highly creative in order to be successful in executing the task.

Therefore, physics education should be taught as part of the basic curriculum in elementary school, junior and senior high school, and first-year university studies. However, the subject content must be suitable for every level, not only for physics majors. It is our challenge to think of new methods of teaching physics and arranging a syllabus, which makes all students, especially girls/women, love physics.

ACKNOWLEDGMENTS

The team members and I would like to take this opportunity to forward our sincere thanks to Ms. Jacquelyn Beamon-Kiene and Erika Ridgway for all their efforts in finding sponsors for us. We would like also to thank Dr. Judy Franz for inviting us to attend the IUPAP International Conference on Women in Physics and to Dr. Marcia Barbosa and all the Conference Committee members for their efforts in making this conference happen. Last but not least, we send our sincere appreciation to UNESCO for their sponsorship.

Iranian Women in Physics

A. Iraji-zad[1], A. Pourghazi[2], and M. Houshiar[3]

[1]Sharif University of Technology; [2]Isfahan University; [3]Shahid Beheshti University

In Iran, women are seen as the managers of families and the trainers of the next generation. The education of women is therefore paramount to the development and progress of the country.

It is well known that physics is one of the foundation subjects of the new world. The general and specialized knowledge of physics is essential to the modern achievements of science and technology, and therefore to the development and progress of the country.

Thus, the education and participation of women in physics is a very important issue in Iran.

The history of science in Iran goes back a long way. We can point to contributions of Iranian scientists in areas such as astronomy, optics, chemistry, geometry, and algebra. However, the actual beginning of teaching physics in its modern form began comparatively late (1920s). Research in physics began 30 years later, and then only on a limited scale.

Systematic education of women in Iran started later than that of men, and on a smaller scale. In the beginning women were mostly interested in the humanities. It took them more than one decade to turn their attention to medicine and the basic sciences, with chemistry and biology being more popular than physics and mathematics.

The social and political changes created by the 1979 Islamic Revolution in Iran have led to an increase in participation of women at various levels of education (Table 1). There are more schools and universities in the country now, making them more accessible to women. It is also more accepted for women to attend these facilities, leading to job opportunities for some women graduates as physics teachers in the girls' schools.

TABLE 1. Participation of Females and Males at Various Levels of Education.

Year	Graduates From High School			Physics Students at Universities									Number of University Physics Departments		
				B.Sc.			M.Sc.			Ph.D.			Up to B.Sc.	Up to M.Sc.	Up to Ph.D.
	F	*M*	*%*	*F*	*M*	*%*	*F*	*M*	*%*	*F*	*M*	*%*			
1985	75,820	83,359	47	1246	3398	27	6	62	9	0	0	0	23	6	0
1990	111,683	136,649	45	1578	5670	22	18	141	11	0	22	0	26	12	3
1995	245,685	199,018	55	3989	6951	37	76	425	15	9	58	13	49	20	6
2000	395,547[a]	316,035[a]	56	10,573	8055	57	127	532	19	30	198	15	63	24	11

[a] 1998 data.

As seen in Table 1, the increase in the number of institutes and women participating at all levels is quite noticeable—the percentage of female undergraduate physics students is now more than 50. But there is a relative decrease in women's participation at the higher levels. The same decreasing trend continues as we move up to the academic and executive levels, as shown in Table 2.

TABLE 2: Participation of Females and Males at Various Academic Ranks and Positions.

Instructor			Assistant Professor			Associate Professor			Professor			Executive Positions		
F	*M*	*%*	*F*	*M*	*%*	*F*	*M*	*%*	*F*	*M*	*%*	*F*	*M*	*%*
46	373	11	15	221	6	1	74	1	0	23	0	5	400	1

CP628, *Women in Physics: The IUPAP International Conference on Women in Physics,* edited by B. K. Hartline and D. Li

The main reasons for the lower participation of women, especially at higher levels of education, lie in the social and cultural structure of Iran:

- As mentioned above, formal education for women started later than that of men.
- The traditional role of women in daily family tasks, and men traditionally being responsible for providing the financial requirements of the family, has made women less inclined to have a job outside the home and, therefore, less motivated to pursue their education.
- For the same reason, fewer resources have been devoted to the education of women, both in the family and in the society.
- As the number of institutes offering positions at higher levels decreases and they are more localized (due to lack of research facilities), those interested may have to leave their families more often. (For example, until recently there were very few Ph.D. programs offered in Iranian universities, and most students had to go abroad for Ph.D. studies.) This creates an obstacle for women, because culturally they are more attached to their families.
- The small number of women in the high ranks of academic, executive, and policy-making positions has denied female students moral encouragement, special regulations for their needs, and suitable policies aimed at helping them through their difficulties.
- Those women who, despite the above-mentioned obstacles, do pursue their education, tend to go less for physics and mathematics because these are considered more suitable for men.

OUTLOOK

The outlook for the participation of women in science in general, and in physics in particular, is bright for the following reasons:

- The average number of children in families has decreased, which has created more free time for women.
- The present economic situation has caused the need for both spouses to work.
- Wider distribution of institutes of higher education has made it easier for women to study without having to leave their families.
- Government is placing a high value on science and education for men as well as women for the development and progress of the country.
- There is a change in the belief that physics is not suited for women.
- The low income of careers available for graduates in physics has caused men to pursue other areas, which has left the scene less competitive for women.

Certainly an encouraging factor is the increase in the last two or three years of the number of women in the higher academic and executive ranks of physics.

RECOMMENDATIONS

1. Create national government and nongovernment organizations in order to reach the following objectives:
   - Create a long-term program to remove the visible and invisible obstacles to women's progress in education, research, and application of physics caused by the country's cultural and social structure.
   - Apply a policy of positive discrimination for women in key positions for a specified period, then henceforth promote on the basis of merit.
   - Increase the participation of women in Ph.D. programs by providing opportunities, advertisements, and encouragement.
   - Develop workshops and training courses in education, research, and management in order to bring women to the forefront of physics.
   - Establish research networks by facilitating communication between active women physicists, requesting assistance from successful male physicists, and providing support to female researchers in finance and equipment.
   - Provide opportunities and encourage women to enter into the levels of decision making, management, and application of physics, and in the research and development sectors of industry.
2. Develop regional and international activities for women physicists to:
   - Improve access to educational and training opportunities in physics for young and talented women graduates at international advanced science centers.
   - Enhance the productivity of women physicists in developing countries through joint projects.
   - Empower women to assume leadership roles in physics and share their views and experiences.

Women in Physics in Ireland: Preliminary Report for the Working Party

Eithne McCabe[1], Aine Allen[2], John O'Brien[3], and Sue McGrath

[1]Trinity College Dublin; [2]Tallaght Institute of Technology; [3]University of Limerick

CAREER ISSUES

A questionnaire was circulated to physics departments nationally requesting input from female staff and postdoctoral fellows. Names of women in physics in industry were also solicited from survey respondents and heads of departments, and questionnaires were sent to these women. The overall response rate to the questionnaires was poor. This could be either a positive or negative sign.

On the positive side, it could mean that women see that there is no real gender issue in physics. On the negative side, it could mean that women are too disillusioned to reply or to voice their concerns. Based on the limited number of survey responses so far, the low response rates may be an indication of both of these aspects. Until a larger response rate has been achieved, it does not make sense to analyze the results statistically, but instead to highlight the overall messages in women's responses.

The very good news is that women in their 20s see essentially no gender issue in the participation of women in physics. They see no potential glass ceiling. Generally women in this age group are very satisfied with their career progression and anticipate no difficulties. The male-female ratios they have experienced as undergraduates are not remarkable, often around 60:40. As graduate students the vast majority of women have had no difficulty finding advisors and report their support from their advisors positively, with comments generally ranging from good to excellent. As postdoctoral researchers, most respondents in their 20s reported no difficulties and were largely positive about their experience.

The responses from women in their 30s are different—some difficulties appear at this stage. This appears to be when women aspire to tenured positions. Women who continue in postdoctoral positions at this stage express considerable dissatisfaction. They can see no well-defined career structure as a nontenured researcher in physics. They see that very few tenured positions in physics are available in Ireland, and although many would like to stay in research, many respondents envision that they will not be able to tolerate indefinitely the poor salary, lack of benefits, and limited opportunities for career progression they experience as a postdoctoral researcher. One respondent mentioned that the only reason she could afford to continue as a postdoctoral researcher was that her partner earned a significantly higher income.

The lack of permanency and of any future of permanent employment is a significant issue for these women. Interestingly, most respondents do not see this as a gender issue.

A significant proportion of women in physics respondents in this age group are not married. Of those who are married, only one woman cited this as a problem. She highlighted the problem of two partners trying to find postdoctoral positions in the same location. The temporary nature of the research contracts and hence the pressure to move geographically on a regular basis proved a stumbling block in this case. This woman abandoned an academic environment for a more economically and geographically stable position in industry.

Surprisingly, most of the women who replied in this age group had no children and seemed unaware of any difficulties that might arise in this context. A very small number of respondents had just one child and mentioned the difficulties of working long hours. On a small sample it is difficult to comment on these findings. It does, however, raise the question as to whether women in physics choose to have children later than previous generations to try to secure a tenured position beforehand, or whether women leave physics around the stage when they start a family because of the time commitments involved. However, these are issues which women face in a wide range of careers. What may make physics different from some other careers is that when one leaves the subject for some years to take a career break—to raise young children, for example—it is extremely difficult to return to a research career when the whole research area will have changed during that time.

Only a few women in their 40s and 50s responded and none older than that did. One response indicated how detrimental working part-time can be in terms of an academic career.

PHYSICS IN SCHOOLS

Most respondents reported access to physics in secondary school. This situation changed dramatically in the 1980s in Ireland when a special initiative to introduce physics to girls' schools was launched by the Irish government. Many schools in Ireland are single sex, and this initiative has been very successful in making physics accessible to girls in recent years. However, many respondents feel that the quality of physics teaching is inadequate and that physics teaching is worse than for other subjects.

WOULD RESPONDENTS CHOOSE PHYSICS AGAIN?

This question received a very mixed response. A significant number of women would not choose physics again as a career, even tenured women. Many were not forthcoming about their now negative perception of physics. Others see physics career opportunities as fairly limited.

GENDER ISSUES IN PHYSICS?

Women see essentially no gender issues in physics early in their careers. By mid- to late-career stages the situation appears to change somewhat if women combine family and career. This is a problem common to many professions. The dedicated full-time career structure with lengthy hours and without career breaks is typical of successful physicists, but may or may not be suitable for both men and women combining work and family.

Women in Physics in Israel

H. Abramowicz[1], S. Beck[1], Y. Shadmi[2], and I. Tserruya[3]

[1]Tel Aviv University; [2]Technion, Israel Institute of Technology; [3]Weizmann Institute of Science

Women are underrepresented in Israeli physics at all levels. Only 25% of the students who choose the physics concentration in high school are girls, and only about 15% of the undergraduate physics majors and 10% of the graduate students at Israeli universities are women. Among senior faculty in physics, who are the role models for the next generation, only 5% are women, and this fraction has not increased significantly in the last decade. But this has only recently and grudgingly begun to be perceived as a "problem" by the Israeli physics community. (While the Israeli parliament has been pushing for a policy of affirmative action in Israeli academia for some time, this has so far been opposed by the universities.)

Although there are no scientific surveys of the subject, it is the usual impression of those in the sciences, both men and women, that women are not discriminated against, but rather that it is cultural factors that keep the numbers of women down. This naturally leads to the attitude that improving the lot of women in physics is a problem for families, for grade school, for high schools—for anyone but the professional physics community! To some extent this is true. Many schoolgirls are simply not exposed to a serious science and math education. Unlike what is reported from other countries, this is a problem in both mixed and single-sex schools. In fact, many all-girl secondary schools do not offer the physics concentration, although it is one of the most common in the all-boys schools. Fewer than one-third of the students in the accelerated high-school mathematics program are girls, and most of them are the children of scientists or engineers. So even before students get to the university, much of the "damage" has been done.

FACTORS PECULIAR TO ISRAEL

The ingrained attitude that physics is "not for girls" is found in many countries, as well as Israel. There are other factors peculiar to Israel that hinder women in academia in general and in physics in particular. Israel is generally perceived as an egalitarian society because both men and women do military service. However, precisely this military service creates a huge difference between women and men. Women do not serve in combat roles, and have traditionally served administrative, rather than technical, posts. Young people are easily molded at this time in their lives and a girl in the Israeli army is more likely to lose contact with the world of science, engineering, and mathematics than is a boy. This is actually changing: more women are serving in technical jobs and as instructors.

Second, Israeli society is by Western standards rather traditional and family-oriented. Even well-educated women will mostly marry and start to have children before 30, and the average family size is three children. Thus, most women become mothers while pursuing their Ph.D. studies, which of course makes it harder to finish one's work and establish oneself as a physicist.

Third, because Israel is a small place, it is an explicit policy that a fresh Ph.D. spend at least three or four years as a postdoc abroad, usually in the U.S. or Europe, to establish oneself before returning to a job in Israeli academia. But if the fresh Ph.D. is a woman, she is probably married and a mother and must face the hardship of taking her family overseas. In these circumstances it is tempting to take another job in Israel and give up on academic advancement.

So women who do study physics in Israel are more likely to look for employment elsewhere than in the universities. This creates "negative feedback," because women students see so few women faculty members that they feel even more isolated and uncertain about whether they would be able to overcome the difficulties of launching a successful career.

CP628, *Women in Physics: The IUPAP International Conference on Women in Physics,* edited by B. K. Hartline and D. Li

As mentioned above, there has been little awareness of the problem in the Israeli physics community, so not much has been done to help women advance in physics. For example, in most American universities and research laboratories, the women faculty and students will meet regularly, in a social setting, for discussions and mutual support. Many women have found this very helpful at all stages of their careers. There is, to our knowledge, no such group meeting anywhere in Israel. However, we have hope that the situation may change: Israel now has a National Council for Women in Science and Technology, the government's Chief Scientist is a woman professor of engineering, and other committees and forums have been established.

Women in Physics in Italy: The Leaky Pipeline

Elisa Molinari[1], Maria Grazia Betti[2], Annalisa Bonfiglio[3], Anna Grazia Mignani[4], and Maria Luigia Paciello[5]

[1]Università di Modena e Reggio E. and INFM National Center on nanoStructures and bioSystems at Surfaces (S^3);
[2]Università di Roma "La Sapienza" and INFM; [3]Università di Cagliari and INFM;
[4]CNR Istituto di Ricerca sulle Onde Elettromagnetiche (IROE);
[5]INFN Sezione di Roma I

Italy is often considered a fortunate country for women in physics. Indeed, the number of women among students in higher education and in the early stages of careers is relatively high, certainly much higher than in most other countries worldwide. However, the percentage of women among physicists decreases very rapidly with increasing level in the profession; also, the presence of women in positions of power is generally negligible.

Undergraduate courses in physics in Italy are now well attended by women, who are generally very successful in their studies. The percentage of women among students in physics has grown from 20.8% in 1960 to 36.4% in 1999 (532 degrees awarded to women and 929 to men in 1999).[1,2] Complete data are not available for the Ph.D. level; however, from our analysis of a few universities we know that a similar percentage is maintained, or even increased, in graduate courses. While the numbers are still far from 50%, they certainly point to less pronounced mechanisms of exclusion of girls and young women with respect to other countries, particularly those that are most advanced technologically, and to a clear similarity with Latin countries such as France or Spain. The debate on the reasons is very interesting and still open, but cannot be addressed in this brief report.

At the entrance to a career in research or higher education we thus find that at least one-third of the possible candidates are women. Let us now follow the situation at different levels. It is not easy to gain a general picture about postdoctoral fellowships, because these are not monitored in official statistical data. However, the Istituto Nazionale per la Fisica della Materia (INFM) reports that postdoctoral fellowships are awarded to women at a rate of between 32% and 43%, depending on the type of fellowship.[2] These data seem to indicate that, at least among condensed-matter physicists, a large fraction of those who enter the very first level of the career with a postdoctoral position are women.

As we proceed along the careers the percentage of women decreases very rapidly. In the university system the portion of women among those holding permanent positions in physics is 15.3%. It is interesting to look at the distribution among the three tenured levels: *ricercatore*, *professore associato*, and *professore ordinario* (approximately corresponding to lecturer, associate professor, and full professor): with increasing level, the percentage of women drops from 25.6 to 15.0 to 4.9 of the total.[3] Note that the situation at the highest level is even worse if one looks at other fields, such as electronic engineering, where physicists are also sometimes employed: here, out of 108 full professors, only two are women.[3]

The trend is similar within public research institutions. There are two national institutes that employ mostly physicists: the Istituto Nazionale di Fisica Nucleare (INFN), focusing on nuclear and high-energy physics, and the Istituto Nazionale per la Fisica della Materia (INFM), focusing on the physics of matter and materials. Here the percentage of women among researchers is 18.7 and 18.4, respectively. At INFN, this percentage decreases rapidly with increasing career level, from 24.1 (level III) to 17.1 (level II) and 4.4 (level I, highest).[4] At INFM the numbers are too small for significant statistics at the highest levels, but the trend appears to be very similar, or worse.[2]

Physics is also present, among other disciplines, at the Consiglio Nazionale delle Ricerche (CNR) and the Ente per le Nuove Tecnologie, l'Energia e l'Ambiente (ENEA). Here the available data aggregate physics with mathematics, and show women in the three levels at 33.7%, 22.8%, and 11.8% at CNR; and 25.2%, 18.4%, and 15.1% at ENEA.[2] In general, it appears that the percentage of women among physicists decreases sharply after the Ph.D. and postdoctoral fellowship levels. The first bottleneck is in the access to permanent positions; at each further step in the career, more and more women are left behind: a very leaky pipeline.

CP628, *Women in Physics: The IUPAP International Conference on Women in Physics*, edited by B. K. Hartline and D. Li

The presence of women in governing bodies, or in general in positions of power in physics, is even more discouraging. Not a single woman is present on the Executive and Directive Boards of INFM and INFN; at CNR one of the seven members of the Directive Board is a woman. None of the presidents of public research institutions is a woman. Even among the directors of local research laboratories, sections (INFN), units (INFM), and institutes (CNR), the presence of women is almost negligible, as it is among the directors of physics departments at universities. We have noticed that in a few cases the choice of women (for participation in boards, committees, etc.) is more likely when the components are elected rather than nominated. In general, however, women in physics in Italy still appear to be excluded almost completely from positions of power and governance. As far as career and power are concerned, the Italian case is thus quite different from, for example, the situation in France.

Amazingly, a public debate on this situation has started only very recently, mostly thanks to the concurrent actions at the European level[5] and internationally, to the publication of official data[1,2], and to the creation of Equal Opportunity Committees (CPOs). CPOs must exist by law at universities and public research institutions; CPOs are now active at INFN, CNR, ENEA and in many universities; a CPO is also being created at INFM. The discussion is now focusing on the reasons behind the present situation and plans for future actions. Among the main issues are:

- *Mechanisms for evaluation of research and teaching (both of individuals and institutions):* It is generally felt that women would especially benefit from an effort to establish more fair and objective evaluation mechanisms. We feel that pressure in this direction would also represent a key contribution to the improvement of the Italian university and research system in general.

- *Mechanisms for selection of people in governing bodies and positions of power:* The main mechanism is still the "old boys' network," which plays against women and often against the most innovative and active scientists in general.

- *Age and mechanisms of access to the first steps in careers and to permanent positions:* Access to permanent positions occurs at a relatively late age (e.g., compared with France), hence maternity can affect the possibility to compete and is sometimes felt as an alternative to career by young women physicists. The most common fellowships do not foresee maternity leaves (such leaves were recently introduced at INFM but not yet at most other institutions and universities).

- *Working conditions and everyday life in labs:* This includes the organization of working time and space, but also the climate for women in the labs (e.g., collaborative vs. competitive/aggressive environments).

- *Role models:* When talking with young women physicists we often feel that very few women represent a model of achieving success in physics together with a "normal" life that includes friends and family and other interests. Role models are also needed to show that success for a woman does not imply adjustment to aggressive behaviors that are more common to men. Women and men may have different definitions of career success. It is important to encourage women to work out role models meeting their requirements and feelings.

All of these are very general issues: more actions, more consultation, more transparency, and more democracy are needed to face power mechanisms that still act against women and other weak groups in Italy.

REFERENCES

1. ISTAT, *Rapporto sull'Italia* (Il Mulino, Bologna, 1999); *Donne all'Università* (Il Mulino, Bologna, 2001).
2. *Figlie di Minerva*, edited by R. Palomba (Franco Angeli, Milano, 2000).
3. Ministero dell'Istruzione, Università, e Ricerca (MIUR) (2001); www.miur.it/.
4. R. Alba et al., *Relazione del Comitato per le Pari Opportunità dell'INFN* (2000); www. lnf.infn.it/cpo.
5. European Commission, *Science Policies in the European Union—Promoting Excellence Through Mainstreaming Gender Equality*, a report from the ETAN (European Technology Assessment Network) Expert Working Group on Women and Science (Office for Official Publications of the European Communities, Luxembourg, 2000); www. cordis.lu/improving/women/documents.htm.

Summary of Nationwide Survey:
The Work Environment of Physicists in Japan

Kazuo Kitahara

International Christian University

This is a report on a cooperative survey conducted in 2001 by the Physical Society of Japan (JPS) and the Japan Society of Applied Physics (JSAP), to study the work environment of physicists. This project was epoch-making in that the survey was not limited to women (contents of the questionnaire are available at www.jps.or.jp and www.jsap.or.jp). We found that the many problems in working conditions often associated specifically with women are actually shared among men as well.

RESULTS OF SURVEY

Differences in working conditions between genders are smaller among the younger generation. In industry, which is more typically represented in JSAP, the ratio of women to men below the age of 40 is higher than in the older generations (Figure 1a). This may be due to social and political progress that resulted from the International

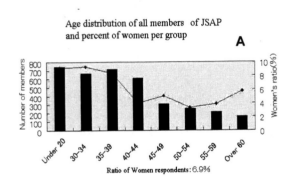

Age distribution of all members of JSAP and percent of women per group

A

Ratio of Women respondents: 6.9%

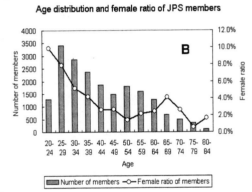

Age distribution and female ratio of JPS members

B

Number of members —O— Female ratio of members

Figure 1

Women's Year of 1975, which resulted in the Equal Employment Opportunity Law of 1986 (EEO). Those who graduated from universities or finished their master's studies in 1985 are now around age 40. Therefore, it appears that the EEO significantly increased the number of women in professional positions, at least in industry.

In academia, however, the employment of women faculty has been more heavily influenced by the policies of the Ministry of Education rather than by the EEO. The science boom of 1966–1970 created many faculty positions and gave opportunities for women scientists who are now in their 60s. In the 1970s, the total number of faculties in universities was fixed, with no prospect of expansion. It became difficult for scientists (now in their 50s) who had just finished their graduate studies at that time to get positions. The next recovery was not until the policies of the "Second Baby Boom Generation" of 1990–1995, which gave rise to many opportunities for women scientists (now around the age of 30) just finishing their graduate studies. Hence, the policies of the Ministry of Education seem more significant than the EEO in affecting the working conditions of women scientists in academia.

Analysis of "position indices" shows significant gender differences in industry as well as in academia (Figures 2a and 2b). In industry, the position of a person, man or woman, becomes higher as his/her age reaches 40 and above. The promotion of women, however, stops at the age of 40. Beyond the age of 50, there is a significant difference in the position indices of men and women. The percentage of men in their 50s (the critical age range for promotion to executive positions) who are promoted is much higher than that for women of the same age group. In industry, however, there has been an increase in the number of women executives in their 40s, probably due to the effect of the EEO and a recognition of the high level of performance demonstrated by women in the past decade.

CP628, *Women in Physics: The IUPAP International Conference on Women in Physics,* edited by B. K. Hartline and D. Li
© 2002 American Institute of Physics 0-7354-0074-1/02/$19.00

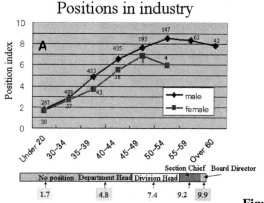

Positions in industry

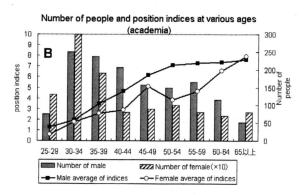

Number of people and position indices at various ages (academia)

Figure 2

On the other hand, in academia, women aged 40–50 are in a difficult situation. Not only is the employment ratio between women and men small, but few women are promoted to higher positions. Women who obtained positions in the Second Baby Boom Generation were promoted more. Thus, we found that the employment situation and success in gaining academic promotions is correlated. That is, when the employment ratio for women of a given age group is poor, their chances for promotion are also poor. This is especially evident for women in their 40s and 50s.

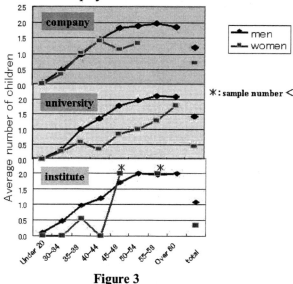

Number of children of physicists in different institutions

Figure 3

We also conducted a survey of men and women for their views on how to balance family needs with career demands. They share common ideals, but it was found that men work more in offices or in laboratories, and less in housekeeping. Both said they look forward to a better work environment to improve their balance of work and family.

Overall, the average number of children is greater for men scientists than for women scientists. However, women who are able to take Child Care Leave, officially introduced in 1992, have a number of children that is much closer to the average figure for men. This implies that official support such as Child Care Leave enables women physicists to raise children more easily. In fact, in industry, there is no gender difference in the number of children up to the age of 44, but beyond 45, there is a difference. In academia, however, the introduction of Child Care Leave has had no significant effect; gender differences still exist in the 30s and below, perhaps because many women scientists that age in academia are still in temporary positions.

CONCLUSIONS

1. More flexible hiring practices are necessary to increase the number of women in academia. In industry, several people are often hired at once. In academia, usually a single post is announced and then filled, leaving less opportunity for women to obtain positions.

2. Current recruiting practices should be reassessed. If we emphasize only academic achievement when hiring, women who devoted the early part of their academic career to raising children may be at a disadvantage.

3. The dependence of promotion on age should be reconsidered. In most cases, those in their 40s are considered most seriously for promotion to full professorship or to executive positions. This, however, can lead to serious conflict between family needs and career demands. If the promotion period can be more flexible, there may be more options for men and women physicists.

The Status of Women in Physics in Kenya

Ruth Luyayi Wabwile, Agneline Yongo, and Sarai Mukhwana Antonate

IMPORTANCE OF HIGHER PARTICIPATION BY WOMEN

Kenya, being a third-world country, has to consider several factors to help realize its national goals. Making the best use of all of its resources, including its human resources, is vital. Increasing the number of women participating in physics would give Kenya more human resources (source of manpower), which will lead to a higher output for economic development.

A greater number of women in physics will lead to:

- Maximization of the talent pool that will lead our country in technological development.
- Hastening of industrialization due to greater available manpower.
- Empowerment of women and liberation from the inferior status they now hold.
- Role models for the younger generation to emulate and thereby do their part toward closing the gender gap in male-dominated careers.

CULTURAL BARRIERS

In Kenya, I have not come across a woman scientist who has made a discovery in physics. Discoveries are made mainly by men. This puts physics in Kenya in a weak position. The major reasons that have made the situation bleak are as follows:

- Most people (men and women) believe that physics is not meant for women. In higher learning institutions, physics as a subject is not taught as much as the other sciences.
- Most women have no time or money to carry out research because they are tied up by family obligations.
- Physics as a subject is marginalized. Though vital, physics is not popularized in our country. Most people, both men and women, think that this subject is only for a few exceptionally bright people (geniuses).

Traditionally, any difficult task is considered a male domain. For a long time, girls have been associated with lighter household chores. Any woman who tried to venture into a career considered masculine was despised by her culture. The general society considers physics a difficult subject; hence it has always been associated with boys. Society still discriminates against women taking certain careers because of male chauvinism. Thus, most girls choose subjects or courses that will prepare them to play their society-given roles effectively. This results in most women taking home economics or courses that prepare them for secretarial and clerical jobs.

Few women will opt to take careers in physics. Culturally, a woman who has ventured into a male-dominated field is looked upon as a social misfit and therefore cannot fit into a domestic environment. This forces a woman to choose between a career and a family. It is believed that such a woman cannot make a good wife.

Professionally, most people have no trust in women. Since they believe women are weak, they also doubt their abilities. Most employers still prefer men, who they think are more effective and reliable than women with the same qualifications. They believe a woman is too tied up by family obligations to perform her duties well.

CP628, *Women in Physics: The IUPAP International Conference on Women in Physics,* edited by B. K. Hartline and D. Li
© 2002 American Institute of Physics 0-7354-0074-1/02/$19.00

REWARDS FOR KENYAN WOMEN

Although they have experienced many hard times in this male-dominated field, women who have chosen careers in physics-related fields have had joys and satisfaction as well:

- *Recognition in society:* Women with careers in physics research, teaching, and administration occupy a high place in society, in which everyone recognizes them and accords them a higher level of respect than their male counterparts.

- *Respect and admiration:* Women physicists are identified by the community as role models to be emulated by the younger generation and sought as a source of advice.

- *Intrinsic and extrinsic motivation:* Women physicists are motivated by even the slightest improvements made in their field, as well as by their own accomplishments.

- *Change of attitude:* Women in general are motivated when another woman presents her research at a conference. This reinforces their belief that what one of them has done, they can do also.

- *Satisfaction:* The most fulfilling and satisfying joy is the realization of personal self-esteem. Since physics careers are male-dominated, women physicists are in a position to lead both men and women, which reverses the traditional role of men leading women.

The Current Status of Korean Women Scientists and Affirmative Actions by the Korean Government

Kwang Hwa Chung

Korea Research Institute for Standards and Science, South Korea

Throughout 5000 years of Korean history, the status of women in Korea was very high. There was no discrimination between daughters and sons, and parents divided their properties equally among all their children. Wives had their own property, and married daughters lived with their family until their children became adults. But the two great wars caused by invasions by Japan and China from the end of 16th century until the middle of 17th century changed Korean society completely. Confucianism became prevalent and a very strong patriarchal system prevailed. Korean women were restricted to the home. When they had to go outside the home, they had to wear clothes that covered their whole body and face, similar to that of an Islamic fundamentalist society. After World War II and the Korean War, the strong influence of the western culture changed the status of Korean women. It has been improving gradually, although at a very slow speed.

Throughout, parents have always had high educational aspirations for their daughters, although sons always had the first priority. Thus, the fact that most of the highly educated women could not have jobs because of the prejudice against them caused many of these women anxieties and problems.

Although the level of education of Korean women is very high, the use of this resource is very low. According to the United Nations Development Program's "Human Development Report 2000," the Human Development Index of Korea is ranked 31 out of 174 nations, while the Gender Empowerment Measure is ranked 63 out of 70. In the government, there are only two women among more than 20 ministers, and 15 congresswomen among 273 members.

GOVERNMENT INITIATIVES

President D.J. Kim has put much emphasis on science and technology, and increased the R&D budget from around 2% to 5% of the total government budget. He is taking very strong measures to make more use of Korea's resource of highly-educated women. Kim follows the assertive actions taken by each ministry, and has made it mandatory for every committee to consist of at least 30% women. Traditionally, science has been considered unfeminine, and therefore not many women scientists are qualified for those committees at this time. Thus, this sudden attention to the female scientists resulted in rather heavy workloads for a small number of qualified women. But, they try very hard to meet all the demands, for they understand these activities are very important to upgrade their status and lead the way for young female scientists.

The Ministry of Science and Technology (MOST) is taking the most active and concrete measures. They are formulating an "Equal Opportunity in Science and Technology" law aimed specifically at improving employment and training opportunities for women scientists. The law will institute affirmative action programs for women scientists. It would also require research institutes and universities that receive government grants to increase the employment of women scientists to 10% of their entire workforce by the year 2003 and to 20% by 2010.

MOST has already increased to 30% the portion of women serving on review panels overseeing government-sponsored science and technology projects. Additionally, the ministry budgeted almost $3.5 million ($US) this year to assist women scientists in their research and training. It recently launched the WISE (Women into Science and Engineering) program designed to provide female secondary school students with an introduction to the world of science, so that promising young students may choose to pursue careers in science. Through the year-long program, young girls have opportunities to meet with established women scientists who will mentor the future scientists one-on-one. Female college students and those in graduate programs are also welcome to participate.

CP628, *Women in Physics: The IUPAP International Conference on Women in Physics,* edited by B. K. Hartline and D. Li
© 2002 American Institute of Physics 0-7354-0074-1/02/$19.00

In order to establish a basis for the gender-specific statistics that can be used in setting up various policies, MOST started a "Data Base for Woman Scientists and Engineers" in 2000. It also established the "Award for Women Scientists and Engineers of the Year," which will go to three women each selected from the categories of scientists, engineers, and promoters.

At this time, although the Korean government is promoting science and technology, a major emphasis is placed on new high technology such as information and telecommunication technology, nanotechnology, and environmental technology; thus, the number of students choosing to major physics is decreasing. Many physicists express great concern about this situation and try to find ways to increase the interest in physics among students.

Girls majoring in physics have always been considered odd, while majoring in biology was acceptable, so there have always been few girls in physics departments. Nowadays, the situation seems to be even more severe, with fewer girls taking physics seriously. Perhaps this is worldwide phenomenon. Since physics is the basis for all modern science and technology, girls majoring in physics can have wider choices in their careers, and the worldwide effort to entice more girls to pursue a career in physics is extremely important.

Survey of Physics in Latvia

Erna Karule

University of Latvia

Latvia is the one of three Baltic states (Estonia, Latvia, and Lithuania) that belonged for 50 years to the former Soviet Union (FSU) and gained independence in 1991. Research in physics in Latvia is carried out in atomic physics, solid-state physics, magnetohydrodynamics, and nuclear physics.

UNDER THE FORMER SOVIET UNION

University Study

In the FSU, university students received a diploma (no degree) after five years. Physicists usually received their diplomas in physics and physics teaching. With a university diploma one could enter doctorantura (three years). At the end of doctorantura one submitted a thesis and received the Candidate of Sciences degree. The second and the highest degree in physics was the degree of Doctor of Sciences in physics and mathematics. A new thesis was required. Most physicists of Latvia received their degrees from the University of Latvia or from institutes of the Academy of Sciences of Latvia. I earned my Doctor of Sciences degree in theoretical atomic physics from the University of St. Petersburg (then Leningrad). This degree was not offered by the University of Latvia in the FSU.

AFTER INDEPENDENCE

University Study

In 1992 (after Latvia regained independence), universities established the bachelor's degree, which one gets after four years of study, and the master's degree, for those who continue their studies in the fifth and sixth year. At the same time the nostrification of degrees was carried out—the equalization of FSU degrees to degrees we now have in Latvia. The Candidate of Sciences degree in physics was changed to a Doctor of Physics degree, and the Doctor of Sciences degree to Dr. Habil. of Physics (similar degrees are in Germany). In 2000 the decision was made to discontinue the Dr. Habil. degree, so now only the Doctor of Physics degree may be obtained in Latvia.

In 2001 in Latvia, one young man received a Doctor of Physics degree in solid-state physics and one young woman and one young man received their Doctors of Physics in atomic physics.

Most physicists today obtain their degrees from the University of Latvia in the capital city of Riga. In the 1960s about 30%–40% of the diplomas from the physics and mathematics department of the University of Latvia went to women. At the end of the 20th Century the number of women declined, but now once more we have about 40% female students of physics. At the same time, women comprise no more than 10% at most research institutes.

Research Institutions

After independence, the structure of science changed. One after another, the institutes of Academy of Sciences were joined to the University of Latvia. There are now three institutes of physics, all belonging to the university:

1. The Institute of Physics (in Salaspils, 20 km from Riga) was the largest institute of physics in Latvia in the FSU. Now only magnetohydrodynamics remains there.

CP628, *Women in Physics: The IUPAP International Conference on Women in Physics,* edited by B. K. Hartline and D. Li

2. The Institute of Solid State Physics (in Riga) is now the largest of the three institutes. The Laboratory of Nuclear Physics in Salaspils now belongs to this institute.

3. The Institute of Atomic Physics and Spectroscopy (in Riga) was established in 1994 when our Laboratory of Theoretical Physics was moved from the Institute of Physics to the university and joined with the Laboratory of Spectroscopy (established in 1967 by late Prof. Dr. Habil. Elza Kraulina).

One does not have teaching obligations in the institutes, but instead of permanent employment now, there is a grant system with no guarantees for a permanent position. Science grants are given for four years; after this time, in most cases, one can get a new grant.

In 2000, 57 science grants were awarded in physics, six of them to groups whose leaders are women. This ratio corresponds to the percentage of women scientists in physics. For most of the scientists of former institutes of the Academy of Sciences, grant money is the only money from our government. With this money one must maintain the infrastructure and pay salaries, and there is very little or nothing left with which to buy new equipment. Except for *Physica Scripta*, donated by Sweden, there are no international physics journals in our libraries. The Science Council has no funding for journals.

The percentage of women working in physics is the highest in the Institute of Atomic Physics and Spectroscopy. In the Institute of Solid State Physics there are some women who lead laboratories. I am alone as the female leader of a group in the Institute of Atomic Physics and Spectroscopy; the other five group leaders are men. At present there are no women heads of laboratories in the Institute of Physics, but there were when the institute was bigger.

There are no women physicists among the members of the Latvian Academy of Sciences, but some are in the biology, chemistry, medicine, and humanitarian sciences. I suppose one of reasons is that most of members of the Academy are current or former directors of institutes. In physics, where most personnel are men, a woman has never been an institute director.

From unofficial information, it is already known that this year the Institute of Physics and the Institute of Atomic Physics and Spectroscopy will be awarded Fifth Framework Programme Centers of Excellence support. Unfortunately, support will be given only for applied physics at these institutes. It seems that basic science merely serves as decoration to win the projects. Winning Fifth Framework Programme projects will not improve even the winner's laboratory equipment or salaries.

FUNDING SCIENCE

Salaries of scientists are much lower than salaries in business. It is very difficult to live on this salary if one is bringing up small children. Therefore, science often does not look attractive to those from the young generation who do not want to leave their native country.

Economics in Latvia are slowly improving, but this is not yet at all helping the situation in science. It is a bit better now for the Institute of Solid State Physics, which in 2000 received support for the next three years by being named a Center of Excellence under the Fifth Framework Programme. Scientists there now have some extra funding available to attend conferences, for example.

In 2001 the Latvian Government allocated 0.21% of its GDP to science, but in 2002 the allocation dropped to only 0.18%. Unfortunately, this tendency is permanent. Money for science is given to the Science Council of Latvia for use as grants. The 12-member Science Council has two women members.

The Commission of Experts in Physics and Astronomy awards the grants in physics and astronomy. Scientists elect the members of this commission from among candidates who are appointed from institutes and supported by the university, or who are nominated by any interested party. There has not yet been a woman on the Commission of Experts in Physics. The secretary of the commission, an appointed position, has always been a woman.

The University of Latvia established Morberg stipends in 2001, given to the best students. Of eight physics students who received these stipends for 2001/2002, five are women: two at the bachelor's level (Ilze Aulika and Biruta Kikuste), two at the master's level (Rita Veilande and Olga Docenko), and one at the doctorate level (Gunta Mazarevicha).

Latvian women scientists do not complain much about the attitudes of male scientists toward them, but all of them, according to their responses to the questionnaires disseminated by the International Women in Physics Conference organizers, would like the situation of physics to be improved as a whole in Latvia.

Women in Physics in Lithuania:
The Problems and Ways to Solve Them

Alicija Kupliauskiene[1], Dalia Satkovskiene[2], and Saule Vingeliene[3]

[1]*State Institute of Theoretical Physics and Astronomy;* [2]*Vilnius University;*
[3]*Secondary School, Vilnius*

The results of studies on women in science in Lithuania parallel those obtained by the American Institute of Physics in 2000.[1] The percentage of women decreases markedly up the academic and career ladders in scientific institutions and universities in Lithuania. Women comprise 20% of the students in the universities, and 16% of all Ph.D. degrees are awarded to women.[2] But the number of women does not exceed 1%–2% in the two top management levels of universities and scientific institutions.

Our analysis shows that the factors influencing the lack of women in physics can be separated into two groups.

The first group of factors are those caused by the gender identity of women and the specific role predetermined for them in nature and in the society.[3,4] In our opinion, these factors are principal. They influence the way of thinking, priorities, and attitudes of women.[5] The relatively long time necessary to take care of children is often fatal to their career in physics. The best time is lost, and it is very difficult to get back the lost positions and necessary qualifications. Physics is a science where the percent of women is relatively small. It looks like the answer to the question of why physics is not as attractive for women as the other sciences is closely connected with both the characteristic features of women's psychology[3] and the peculiarities of the scientific field,[5] and that these factors should be seriously studied.

The second group of factors depends on the local environment and includes the economic situation in the country, its traditions, the existing system of child care and education, the prestige of the profession, and more. Society in Lithuania is used to seeing women doctors, teachers, nurses, and saleswomen. It is no wonder that the image of a woman is formed as a representative of these professions. Even a brief study of school textbooks shows that men and boys are depicted more often than women and girls. The perceptions and prejudices of the teachers are formed not only by traditions. Boys are learning technologies at school when girls are trained in housekeeping. Boys are gaining experience and knowledge in experiments and measurements, therefore the connection of a simple electric circuit or use of a caliper do not cause a problem for boys, whereas they are big problems to girls because they are doing them for the first time. Naturally, the girls look like gawks. Under the pressure of traditional thinking, talented girls choose to study physics in a teacher-training college or pedagogical university, or choose more memory-oriented sciences like biology, medicine, or law in universities. In these specialties the numbers of girls exceed 50%.

It seems that at present the main problem affecting women in physics in Lithuania is the lack of money for research and teaching, and the low prestige of the profession. We need to stress in the media the important achievements in the physical sciences that have benefited society. We need to investigate and discuss the problem of women in science on various levels (from schools up to the government of institutions) and work out a national program to encourage young, talented women to seek careers in science, especially physics.

It is evident that until women are fully represented at the leadership level of public, professional, and economic life, we cannot say that women enjoy full equal rights. On the other hand, if not reversed, this situation will have serious consequences for the future as all countries become more dependent on the availability of a large, scientifically trained workforce. Moreover, gender equality in science will give a new perspective to science itself, physics included.

REFERENCES

1. R. Ivie and K. Stowe, "Women in Physics, 2000," AIP Publ. No. R-430 (American Institute of Physics, College Park, MD, June 2000); www.aip.org/statistics/trends/highlite/women/women.htm.
2. E. Makariuniene and L. Klimka. *Reference Book of Physicists and Astronomers of Lithuania* (Institute of Physics, Vilnius, Lithuania, 2001).
3. S. M. Burn, *The Social Psychology of Gender* (McGraw-Hill, Inc., New York, 1996).
4. M.A. Pavilionienë, *Drama of Sexes* (Vilnius University Press,Vilnius, Lithuania, 1998) (in Lithuanian).
5. E.M. Byrne, *Women in Science: The Snark Syndrome* (The Falmer Press, London, 1993).

Women and Physics in Malaysia

Khalijah Mohd Salleh[1], Azni Zain Ahmed[2], and Samirah Abdul Rahman[2]

[1]Universiti Kebangsaan Malaysia; [2]Universiti Teknologi MARA

In Malaysia, women are generally well respected and given equal chances with men to pursue their careers. Women holding high positions in society is rather common. The governor of Malaysia's national bank and Malaysia's former attorney general are among the top women achievers in Malaysia. The success of women's participation is due largely to the equal educational opportunities and access to educational facilities funded by the government.

Government spending on education in Malaysia is at par with the developed countries, which currently stands at 5.2% of the GNP, as compared with Australia (5.5%) and the United Kingdom (5.3%). Support for women was given another hefty push in late 1999 when the Women's Affairs Department was moved to the Prime Minister's Department. The Ministry of Women and Family Development was created and headed by a woman minister, which firmly established the seriousness of the government about women's issues.

Part of the strategic thrust for the advancement of women includes increasing the female population in the labor market; providing more education and training opportunities for women to meet the demands of the knowledge-based economy and improve their upward mobility in the labor market; reviewing laws and regulations that inhibit the advancement of women; improving the health status of women; strengthening research activities to increase participation of women in development and enhance their well-being; and strengthening the national machinery and the institutional capacity for the advancement of women.

Malaysia's Vision 2020 is a grand plan to turn the country into a fully industrialized economy. A knowledge-based economy was identified as one of the keys to sustaining the country's development. Science and technology, research and development are prerequisites for knowledge creation and innovation.

With this background, the women of Malaysia can and will play an equal role with men to realize the nation's vision. Able women must be made aware of their potential contribution to national development and its sustainability.

PHYSICS AND SOCIETY

Physics is external to the local culture. It belongs to the elitist and is foreign to the ordinary person. Physicists organize meetings that are very technical in nature. By default, non-technical, social, and educational matters are rarely discussed. Thus, the problems of physics being abstract, difficult, and boring for the young and old alike are cause of concern for only a few discerning physicists in the country. These problems continue to exist and may contribute to the declining interest in physics among the young. To make matters worse, lack of physics graduates has given the task of teaching physics in schools to nonqualified teachers.

Physics is not a compulsory subject in the upper secondary level. Physics has been dominated by the males. Although girls are given equal access to science education, they do not choose to study physics unless they are interested in the subject, or when physics is a prerequisite for an advanced study, like engineering or medicine. Those who choose to study physics try their best to do well in their examinations.

However, for the past couple of years Malaysia's greatest concern in the field of education is the drastic decline in the number of students studying science at institutions of higher learning. Thus, the main issue is not so much whether there are girls studying physics but whether there are enough pupils taking up science at schools and later at institutions of higher learning.

CP628, *Women in Physics: The IUPAP International Conference on Women in Physics,* edited by B. K. Hartline and D. Li
© 2002 American Institute of Physics 0-7354-0074-1/02/$19.00

WOMEN IN PHYSICS

Women have been actively involved in research and development activities in government research institutions, institutions of higher learning, and the private sector during the last five years. They make up one-third of the total number of researchers with bachelor's and master's degrees. However, women researchers with Ph.D.'s account for only 24% of the total number of researchers with the same qualifications. Women contribute mainly in the areas of medicine, health, and information services.

Recent surveys show that a substantial number of women physicists exists only in two major research institutions in the country—the Malaysia Standards Institute (SIRIM) and the Malaysia Institute of Nuclear Technology (MINT). There women physics graduates constitute 5.9% and 3.6% of the research workforce, respectively. In SIRIM, however, there are women physicists who have pursued administrative posts such as managers (2.6% of the managerial workforce) and executives (0.9% of the executive workforce). These figures are expected to go down in years to come in accordance with the declining number of science graduates mentioned earlier.

It should be noted that most women who graduated from the 1980s to the early 1990s, graduated in physics from overseas. During those years, the Malaysian goverment offered overseas scholarships to good students specifically to pursue degrees in basic sciences as part of the future planning of the country at that time. When the country suffered an economic downfall and stopped sending students overseas, the number of science graduates fell as well.

In general, very few women physics graduates pursue careers as researchers in research institutions. This is because such careers require very long hours, which is not always possible after the women are married and have additional family commitments and responsibilities.

Although female students at school level generally outshine the males, physics does not seem to be the choice of females at the tertiary level. The total number of women who graduated in physics from Malaysian universities during the last 10 years is only about one-third of the total number of graduates in physics. A recent survey has revealed that the trend is similar in all the universities. Although the number of physics graduates since 1990 has increased fivefold, the percentage of women graduates remains about the same.

A closer look at the type of physics courses and the numbers reveals that more women than men graduate with a degree in physics education, although in general there are more men than women in physics. At the other extreme, fewer than 10% of industrial physics graduates are women. The field that seems to be non-gender-biased is physics with computer science, which attracts an equal number of men and women.

During the last 10 years there has been a shift of interest from the pure sciences to the applied sciences, engineering, and technology. For example, one university produced more than 90% of physics graduates with major in industrial physics. The more conventional physics major of material science managed only a meager 3%. The dismal performance of pure or traditional physics is due to the popularity of applied science and computer science.

A smaller number of women return to the ivory tower or continue to seek a postgraduate degree. If the percentage of women graduates with a bachelor's degree is 30%, the percentages of women graduates with a master's and Ph.D.'s are 24% and 20%, respectively. The discrepancy between women graduates and men is even bigger as the degrees become higher. At the doctoral level, the number of women graduates becomes even more critical. Overall, the percentage of graduates with a postgraduate degree is only 10%, and out of those, only 2% hold a Ph.D. Of the Ph.D. holders, only a quarter are women.

Women who choose to study physics and work in physics-related areas such as research, education, and industry appear to have certain characteristics. They may be soft spoken and demure but are more often women of rather strong personality, vocal, and prepared to face challenges. In terms of career development, there are a handful of women who progressed to the management level, but the number is small compared with the men (1.8% females, 4.0% males). A point to note is that scientists in general, whether male or female, have fewer opportunities to go into the management level than those in the professional fields and humanities.

THE FUTURE OF WOMEN IN MALAYSIA

There is still so much to be done. The country's target to produce a ratio of 60:40 for science and non-science graduates has not been met. Our target for 1000 scientists and researchers per 1 million of population has also not been met. The government of Malaysia has outlined several strategic plans to intensify research and development in science and technology, enhance human resource development, and increase focus on education and training, teacher development, and revamping the education system. To accelerate these efforts, women physicists in Malaysia must also rise, unite, and create awareness among themselves on the current state of affairs and become the catalysts in society toward the betterment of women in physics.

Possible Strategies for Improving
the Situation of Women in Physics in Mexico

Lilia Meza-Montes[1] and Ana María Cetto[2]

[1]Instituto de Física de la Universidad Autónoma de Puebla
[2]Instituto de Física de la Universidad Nacional Autónoma de México

A recent report of the National Commission on Women ("Conmujer") in Mexico, draws attention to the crucial age at which the gap between genders in education starts: at 11 years the highest percentage of women abandon school.[1] In contrast, at higher educational levels female desertion becomes smaller than that of the male student population. Thus, in all graduate programs 41% are female students, the figures being considerably lower in the natural sciences.

The National System of Researchers (SNI), a governmental institution that grants fellowships according to individual accomplishments in scientific work, includes 28% female members, with only 14% in the fields of physics and mathematics. Certainly, the issue is not just one of numbers; it is one of opportunities and of the wealth of potential that women provide. As in other fields of human activity, an increased female participation in science should make a difference not just for women but also for science.[2]

According to the Ibero-American Catalogue of Programs and Human Resources,[3] there are 47 research and educational centers in Mexico, a number of which offer undergraduate and graduate programs in physics or related areas. Barely 9% of the faculty members are women. Of these approximately 200 women, 15 have bachelor's degrees, 53 have master's degrees, and 132 have Ph.D. degrees. The general trend in the geographical distribution is neatly reflected: the scientific community is concentrated in the capital, with 117 of the women working in Mexico City. On the other hand, 38% of the centers include only one or no woman physicist at any level.

The figures are definitely low; however, the proportion is more favorable than in some other countries. Two factors may contribute to this: first, some fields are relatively new in our country. Historical facts, such as political instability and lack of institutional interest in science, hindered the development of scientific centers in the past. Physics as a professional career began when the first Institute of Physics was founded in 1938 and one year later the first school was created, both in the Universidad Nacional Autónoma de México, in Mexico City. The first attempt in the provinces took place in 1950. This means that physics has developed within a more contemporary, liberal environment, with less sexist bias and preconceptions than are found in countries with a longer tradition.

Another possible factor may be our educational system. Most of the high school programs include science courses, which expose students to science even when they do not take it as a major field. This fact allows women to test their capabilities and discover that they are able to follow a scientific career. Of course, a deeper analysis is necessary, but these reflections can give some insight on the cultural differences in the participation of women in physics.

Recently, some efforts have been made to confront gender inequality in science and find ways to overcome them. Since 1990, the bulletin *Supercuerdas* ("Superstrings") for women in science, published in Mexico, has served as a means to establish contact with colleagues from other institutions in the country and in Latin America, and as a forum for the analysis and debate of problems encountered by women in science.[4]

Strategies have been suggested to increase the participation of women in scientific jobs, such as guaranteeing equal opportunities in education, providing professional orientation, and reviewing how work is evaluated to provide equality. Campaigns to modify traditional roles have been developed by Conmujer and several other groups. Furthermore, institutional actions are in progress. SNI initiated an extension of the membership period by an additional year for scientists who become pregnant, as a recognition that this circumstance has a negative impact on their productivity.

CP628, *Women in Physics: The IUPAP International Conference on Women in Physics,* edited by B. K. Hartline and D. Li
© 2002 American Institute of Physics 0-7354-0074-1/02/$19.00

Nevertheless, we have to work out several aspects, some of which are closely related to our specific circumstances, particularly the social situation. Some of the issues were discussed at length at a recent regional meeting held in Bariloche, Argentina.[5] Let us emphasize some of them:

- Lack of support to physicists with children keeps them in a marginal situation.
- Young students do not have access to child care services, and very often this reduces their possibility of fulfilling requirements to obtain a fellowship.
- The attitude of our male colleagues is still sexist.
- Very importantly, the way science is taught in elementary and high school and the biased expectations of schoolteachers and parents towards girls vs. boys, are seen to negatively influence the female choice of scientific careers.

Stimulated by the present Conference, small working groups of colleague physicists and students have been formed in the University of Puebla to work on these and related issues. Conferences like this one give us an opportunity to exchange experiences and collectively try to find ways to improve the situation. As an example of the importance of such actions, we recall the five regional preparatory UNESCO meetings (one of them being the Latin American meeting held in Bariloche[5]) that produced a significant input for the World Conference on Science, which is now reflected in the commitments of governments and institutions contained in the Conference documents.[6] It is very important to organize events such as this one because a significant portion of us carry out our activities in a gender-imbalanced environment. Panels with male participation on topics such as the ones considered in this Conference may make our colleagues more sensitive to these issues, reflect on them, and change their attitudes.

International campaigns like the ones organized by UNESCO and UNIFEM on gender equality (e.g., the contest, "Children's Views on Science in the XXI Century") have an influence on the population. They can be more specifically devoted to attract young females and promoting women scientists, and reinforce the local activities carried out by female scientists. Activities at the grassroots level to address the disadvantages met by women locally, are essential. But joint efforts, recommendations to governments, and monitoring are always helpful; international actions give strong support to national activities.

REFERENCES

1. Programa Nacional de la Mujer-SEGOB, México; http://www.datasys.com/conmujer.
2. A.M. Cetto. Papel de la mujer en la transformación de las ciencias: una óptica diferente. Reprinted in Supercuerdas *11-12*, 9 (2001).
3. *Catálogo Iberoamericano de Programas y Recursos Humanos en Física 2001-2002* (Sociedad Mexicana de Física-FeLaSoFi, México, 2002). Catalog includes data on the major bachelor and graduate programs and research centers in Latin America and Spain.
4. Supercuerdas *1-12* (1990-2001); http://www.arce0.fciencias.unam/supercuerdas.html. Independent bulletin that serves the Latin American branch of the Third World Organization for Women in Science (TWOWS).
5. "Femmes, science et technology en Amérique latine: diagnostic et stratégies," in *Femmes, Science et Technology Vers un Nouveau Développement?* pp. 19-24 (UNESCO, Paris, 1999).
6. "Science for the Twenty-First Century—A New Commitment," The UNESCO-ICSU World Conference on Science; http://www.unesco.science.org/science/wcs.

Women in Physics in the Netherlands

Hélène van Pinxteren[1], Annalisa Fasolino[2], Eddy Lingeman[3], and Christa Hooijer[1]

[1]Foundation for Fundamental Research on Matter (FOM); [2]University of Nijmegen;
[3]National Institute for Nuclear Physics and High Energy Physics

SITUATION IN 2000

The Foundation for Fundamental Research on Matter (FOM) promotes, coordinates, and finances fundamental physics research in the Netherlands. FOM employs about 325 Ph.D. students and 100 postdocs. They work at FOM research institutes and in university laboratories. The foundation also finances and operates a number of research facilities at and in collaboration with universities. FOM operates under the auspices of the independent Netherlands organization for Scientific Research (NWO), which encompasses all fields of scholarship and acts as the national research council in the Netherlands.

FOM estimates that since 1990, the percentage of women obtaining a first-level degree in physics (i.e., a master's degree) has hovered around 6%. The percentage of women obtaining a Ph.D. degree in physics is significantly higher, around 15%. And the percentage of female postdocs is even higher, at 20%. However, this reverse-pipeline effect is not due to Dutch first-degree graduates staying on in physics research. Rather, it is due to a large influx of foreigners who obtain their Ph.D. here or come here to do a postdoc. Around 39% of all FOM Ph.D. students and 72% of the postdocs are foreigners (it should be mentioned that the postdoc positions are in principle not meant for Dutch Ph.D.'s). For women these numbers are even higher: 50% of the female Ph.D. students and 80% of the female postdocs are foreign.

The percentage of permanent female staff members has increased significantly over the past five years, and is now at 3%. Foreign women make up a major portion of this increase. Table 1 summarizes these numbers. We only have direct access to the data of the FOM personnel, but these are quite representative of the overall situation: about 40% of all Ph.D. students and postdocs in the Netherlands are employed by FOM.

TABLE 1. Women Students and Staff at FOM and Universities in 1995 and 2000.

Type of Position	1995 Number	%	2000 Number	%
Ph.D. students (FOM)	36	10	52	16
Postdocs (FOM)	12	13	25	19
Permanent staff (FOM)	1	1	3	3
Permanent staff (universities)	9	*	15	*

* We do not know at present the total number of male permanent staff at Dutch universities.

MEASURES

Since the 1980s, there has been awareness in the Netherlands of the underrepresentation of women in science. Campaigns have aimed at high school students to attract more girls to study science at the universities, but the influx of female students has remained around 12% for almost two decades. The percentage of female graduates is much lower, at 6%; we intend to investigate this effect further.

In 1997, the Dutch government launched a proportional representation act. It is designed to maintain the relative number of women at each step up the academic career ladder. The main idea is that the relative number of women on the next step of the ladder should be at least as high as at the previous step. Because this is clearly not the case for the step from postdoc to permanent staff member, the Dutch funding agencies for scientific

CP628, *Women in Physics: The IUPAP International Conference on Women in Physics,* edited by B. K. Hartline and D. Li
© 2002 American Institute of Physics 0-7354-0074-1/02/$19.00

research as well as the universities have started some programs to improve this situation. We will only discuss here the program of the funding agencies, FOM and NWO.

FOm/f Program

The FOm/f program (m/f stands for male/female) was started in 1999 with the intention of stimulating the participation of women in physics in the Netherlands. The program has three different types of actions:

1. Salary costs: Universities wishing to appoint a woman to a position that will become vacant in a few years (e.g., due to retirement) can apply for salary costs. Through this program five women were appointed to permanent staff positions, and in 2001/2002 two more will be appointed.

2. Research funding:

   - Projects with a female principal investigator (PI) that end up just below the funding cutoff in a regular grant procedure are granted funds from this program. Funds can only be granted once to a female PI. So far, five projects have been funded.

   - Personal postdoc positions: women can apply for a three-year grant to support a position in the Netherlands. This should be combined with a postdoc abroad, financed elsewhere. (A postdoc abroad significantly increases the chances of obtaining a permanent position in the Netherlands.) Two positions have been granted thus far.

3. Awareness measures:

   - In 2000 the Minerva Prize was granted for the first time. This is a biannual €5,000-prize awarded to the best publication by a woman in the past two years. The paper has to be written about physics research done in the Netherlands. The prize is intended to highlight the contribution of women to physics and to encourage women to pursue a career in research.

   - In 2001 the FOm/f symposium was organized for the first time; 105 women were present. The day consisted of scientific talks, two workshops (one on career planning and one on research funding), and a dinner. This day served to bring together women in physics research (no men were present), to decrease the sense of isolation that many women in physics feel, and to establish the foundations for a Dutch network of female physicists. In principle this event will be organized once every two years.

The FOm/f programme is directed only toward women already in physics research, and is not geared toward getting more women to study physics.

Actions Within NWO

Within NWO, two other programs have been started. The ASPASIA program is aimed at getting more women promoted from the assistant professor to the associate professor level. NWO will finance an extra Ph.D. student position if the university where a woman is employed agrees to promote her to the associate professor level. The other program, MEERVOUD, aims to appoint more women to tenured (assistant professor) positions. This measure is similar to the first type of support in the FOm/f program. After a maximum of five years, the university has to bear the salary costs in full.

Women in Physical Science in New Zealand[1]

Jenni Adams[1], Pat Langhorne[2], Eleanor Howick[3], and Esther Haines[4]

[1]University of Canterbury; [2]University of Otago; [3]Industrial Research Ltd.; [4]University Chemical Laboratory

The population of New Zealand is 3.8 million with 85% of its people concentrated in urban areas. Women make up 50.9% of the total population. The labor force draws from 50% of the population. Women in New Zealand now make up close to half of this (45%). However, the numbers at higher levels of career structures are reduced in almost every area or discipline so that even in professions where women are in the majority, men hold the senior positions. The average weekly wage for females is 80% that of males.

PARTICIPATION IN PHYSICAL SCIENCES

School: Although the national curriculum document, "Science in the New Zealand Curriculum" stresses the importance of a high level of scientific literacy for all New Zealand students, less than half of students in senior school study any science subject. Girls show a definite preference for the natural sciences rather than the physical sciences. In 1997 38% of girls chose to study biology in their final year at school, while 17% chose to study physics. For boys the trend was reversed, with 27% studying biology and 35% studying physics.

University: There are eight universities in New Zealand, 23 polytechnics, and many other privately funded tertiary institutions. 193 postgraduate and graduate qualifications were completed in the physical sciences in 2000. The overall percentage of women taking these awards was 18%. For the bachelor's degree, the percentage of females completing was 12%, in masters and honours 29%, and in doctorates 17%.

Employment as physicists: It is difficult to obtain statistics on the number of people working as physicists outside the universities. The total number of employed scientists and engineers in 1996 was 48,546. Of these, 18.9% were women. The best representation of women was as health professionals (35.4%) and the least representation was as engineers (7%). In 2001 there were 66 physicists employed in permanent positions at New Zealand universities, eight of whom were women (12%). The academic hierarchy has four categories (in descending order): professor, associate professor, senior lecturer, and lecturer. Of the eight women in permanent positions, three were lecturers and five were senior lecturers. In comparison, there were 14 men professors.

TWO SEPARATE ISSUES

We feel there are two separate issues: (1) the underrepresentation of females in physical sciences in school and university undergraduate programs and (2) the underrepresentation of women in high-level positions in all professions. Of course, the low numbers of women participating in physical sciences from the onset results in a lower number holding high-level positions in the academic or industry hierarchy when compared with other professions. These issues are not peculiar to New Zealand.

Underrepresentation of Females in Physical Sciences at School and University

It has been recognized that neither boys nor girls have a positive attitude toward physics at school. One study found that boys had, in fact, only a slightly more positive attitude to physics than did girls. The boys, however, accepted that their career options were greater if they studied physics. Therefore it seems there are two points to address: (1) encouraging the perception that physics is an interesting and enjoyable subject (then ensuring that it is!) and (2) educating girls about career options that are opened by taking physics.

---

[1]A full report with references is available from www.phys.canterbury.ac.nz/~physjaa/ women_in_physics.

CP628, *Women in Physics: The IUPAP International Conference on Women in Physics,* edited by B. K. Hartline and D. Li

As an example how the first point may be addressed, we present the profile of a teacher who loves teaching physics and whose students love taking physics.

A GREAT PHYSICS TEACHER: ANNA COX

I discovered that I had an overwhelming interest in physics when I first took the subject at school. Many of the concepts were not new to me. I had already been thinking over the ideas and applications when I pulled my toys, parts of the house, and our broken stereo apart. The mixture of empirical evidence and theoretical derivations presented in the classroom were a more elegant and exacting approach to what seemed like a whole universe of ideas, phenomena, and quirky little unsolvable problems.

My other growing passions of rock climbing, skiing, and surfing all received an enormous boost from a greater understanding of the world according to Newton. It is through a careful combination of the applications of physics and empirical and theoretical development of concepts that I try to introduce this wonderful body of knowledge to my students.

I teach physics at King's High School in Dunedin. Many of the boys are keen surfers and know much about mechanics and wave motion from their own experience of the learning curve of this most complex physics-based sport. Many of the students are serious "petrol heads" and know the inner workings of their car better than the average mechanic. They are quick to gain an appreciation of current electricity, having tinkered lovingly with their cars' circuits.

The keen mathematicians who love a complex proof in their calculus classes are even more thrilled to see the purpose of the exercise when the gradient of a graph, or the area beneath it, is a real and meaningful quantity. The most usual passion of a boy at King's is how to become the first XV's first five eighth, and so how to become a goal kicking machine. They pay particular attention to the practicals and theory covered in the 2-D motion topic!

Every student has a different frame of reference, and can develop a love of this subject so easily. The added attraction of careers in engineering, architecture, sports science, etc., etc., is ample justification for the hard work they put into studying physics. I love teaching physics.

The second point is addressed by one of the New Zealand universities by employing a Women in Science and Engineering (WISE) Equity Advisor. This role was created about 10 years ago in response to the low levels of participation of women in the physical sciences and engineering. The WISE advisor works with the science and engineering faculties to develop initiatives to increase the recruitment of women into science and engineering courses where they are currently underrepresented, such as physics.

Underrepresentation of Women in High-Level Positions Across All Disciplines

The most obvious barriers are the time associated with childbirth and childcare. Since 1987 women in New Zealand have been entitled to one year's leave without pay on the birth of a child. This year legislation was passed introducing paid parental leave for 12 weeks for working mothers. A milestone in itself, it does not address the issues affecting promotion of time away from the workplace and part-time employment. Child-care facilities are readily available and although they may not be economically feasible for women on low incomes, they are increasingly used by professional women.

It also seems that women are unwilling to move into positions that they perceive as more stressful. Increased pay and power are less likely to be motivating factors for women to move up the career structure and are not seen as compensation for the longer working hours and more stressful environment required. We actually found that some men would prefer not to be working full-time but feel forced into this role by society's expectations and by partners who have chosen not to work while rearing children. The long-hours culture is not healthy for men or women—would the presence of women in higher positions help to dismantle it?

One issue that arises due to New Zealand's relative isolation is that many promising young women are encouraged to seek postgraduate and postdoctoral experience in Europe or the United States, which is in itself highly desirable. However, these women often marry overseas and are even less likely than New Zealand men in the same situation to return to New Zealand. This "brain-drain" of both sexes is a major concern for the New Zealand government.

Nigerian Women In Physics

Ibiyinka A. Fuwape[1] and Oyebola Popoola[2]

[1]Federal University of Technology; [2]University of Ibadan

More than 60% of the 100 million people in Nigeria are women. The status of women in physics is affected by the literacy level in different parts of the country, as well as the availability of facilities for teaching science subjects at both primary and secondary levels. Most of the women in the physics are from the southern part of Nigeria. This is expected because the majority of women in the northern part of the country are not educated.

A survey was conducted with the aid of a structured questionnaire developed by the IUPAP Working Group on Women in Physics. Fifty questionnaires were distributed to physics departments in Nigerian universities, but there were only 20 respondents. The returned questionnaires were collated and analyzed.

The study showed that there is a low participation of women in physics. The enrollment of female students is low, as shown in Figures 1 and 2. Only 6% of physicists in Nigerian Universities are women (Figure 3).

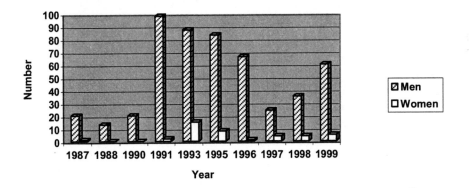

Figure 1. Physics undergraduate enrollment at the Federal University of Technology in Akure for selected years.

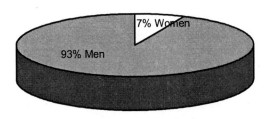

Figure 2. Physics enrollment in Nigerian universities, 1987–1999.

The participation of women in physics is affected by poor exposure to science subjects at the secondary-school level. Very few girls enroll in physics or mathematics because the subjects are perceived to be abstract and masculine. Physical science subjects are not usually portrayed to be vital for daily living. Physics is poorly taught in the secondary and higher schools.

The survey revealed that most of the women in physics developed their interest in the subject while they were in the secondary school. All the respondents indicated that their parents encouraged them to study science subjects. Some of the respondents also reported that their teachers supported their interest in physics and other science subjects, despite limited laboratory facilities.

CP628, *Women in Physics: The IUPAP International Conference on Women in Physics,* edited by B. K. Hartline and D. Li

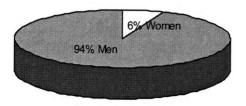

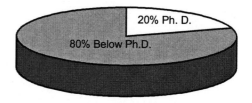

Figure 3. Physicists in Nigerian Universities.   **Figure 4.** Women physicists with Ph.D.'s.

EDUCATION

Most women in physics were one of very few female undergraduate students while they were in the university. Of 546 students admitted to the physics department at the Federal University of Technology in Akure between 1987 and 1999, only 40 were female. The enrollment of female students in the other 26 Federal Universities in Nigeria is similar to that of Akure (Figure 2).

Some of the respondents indicated that they received the same amount of attention from their lecturers, while a few reported that they received more attention than the other students. There were also reports that women in physics sometimes face sexual harassment from their professors or people who are in position to offer assistance.

Almost all respondents did their graduate studies in universities in Nigeria. There were complaints of inadequate laboratory facilities, obsolete and faulty equipment, and poor library facilities, electricity supply, and internet facilities. They indicated that these problems slowed down the pace of their research work.

Only 20% of Nigerian women in physics departments have a Ph.D. degree (Figure 4). Most of the respondents have not been exposed to international research opportunities. Almost all said they work in isolation. All were schooled in Nigeria, hence most of them do not have any contact with the international community.

CAREER OPPORTUNITIES

All respondents are employed as lecturers in universities in Nigeria. There are reports of very few women in physics employed in the research institutes or by the Ministry of Science and Technology. Some respondents reported that they have progressed more slowly in their career than their male counterparts. There were also reports of several women who abandoned physics after their bachelor's degree for more lucrative jobs in banks and finance.

All the respondents are married. Most of them got married after their first degree. Almost all reported that they struggle to combine their career in physics with their marriage and family. More than 80% indicated that their marriage and raising of children affected progress in their research work. Some complained that it was very difficult to combine child care with the rigors of research work, participation at conferences, and international research visits.

RECOMMENDATIONS

1. Efforts should be made by the Nigerian government to improve the teaching of physics beginning at the high-school level by equipping the laboratories and providing teachers on-the-job training. Physics teachers should be encouraged to stay on the job.

2. Girls should be encouraged and not discouraged to study physics.

3. Grants for postgraduate studies abroad, especially in developed countries, should be made readily available to women. A collaboration program could be put in place that will enable a woman in physics in Nigeria to visit research centers in developed countries for 3 months, 6 months, 9 months, 1 year, or more.

4. Training and discussion forums on balancing family and career should be provided for women in physics.

5. Since the career prospects for women in physics is limited to teaching in high schools and universities, many women who would have studied physics prefer to study computer science or other engineering courses where they can get more lucrative jobs. There is a need to provide creative opportunities for women in physics as an incentive for them to stay in the field.

On Women in Norway in the Physics Power Struggle

Åshild Fredriksen

University of Tromsø

What do we mean by leadership and physics power structure, and how is it accessed?

In my view, there are different power structures as well as different levels of leadership. Boards and committees within the department and faculty are elected among their colleagues and may belong to the power structure, but these positions are usually quite laborious with few rewards (at least in Norway). Administrative duties usually come as part of any type of leadership, and hence most acknowledged physicists and those working toward a full professorship do not waste their time on leadership positions not within their direct research interests. For instance, faculty leadership is mainly concerned with serving the entire department on a general basis, and not one's own specific interests.

On the other hand, being the head of a program committee of an international conference or research collaboration, an editor of a renowned journal, or a well-paid director of a research institute is considered an honor and an asset on one's CV. Participation in international research networks and/or programs giving access to healthy funding is highly valued and also eases access to program boards and committees. It is my understanding that these activities are examples of participation in the true physics power structure.

Launching projects to obtain funding for new experiments seems harder for female than male experimentalists. Men seem more daring in initiating and building big- and medium-budget projects than women, and they seem to receive less punishment if they fail. In the long run, I think this kind of activity is important in order to build a research group around one's own interests, and in turn get higher visibility, which is mandatory in order to join the power structure in international physics.

To summarize, there are leadership positions that are prestigious to a physics career and those that are not. The prestigious positions are often obtained through network building and visibility. Visibility is underpinned by being perceived as a good physicist in the physics community. However, being a highly qualified physicist is not the only requirement for obtaining visibility. It is very often accomplished through active mentoring by a former supervisor or some other highly visible senior. Having this kind of mentor guarantees a flying start to gaining successful access to the physics power structure, and probably also gives a boost to self-confidence that enables one to take on leadership initiatives and tasks.

Women very seldom receive this kind of mentoring, no matter how well they match the skills of the male physicists who are supported this way. There may be many reasons for this: if a woman competes directly with a man, the mentor might perceive the man's career more important than the woman's. Often the ambitions of a young woman are underestimated, especially by the woman herself. She can easily be perceived as less ambitious because of her more modest appearance. It may even be enough if she has been heard talking negatively about being totally absorbed by work, or expressing a wish to have children and a private and social life besides physics. Male (and female?) mentors are more than willing to take a woman's statements literally and write off any ambitions on her part, which also reinforces the general notion that women are not ambitious. The result may be that her research goes unnoticed by the physics community, while her male colleagues quickly enter project/program committees and get invited to give talks.

What other reasons can keep women out of the power structure?

Lack of self-confidence is one of the reasons. This will often refrain female physicists from making contacts and participating in entrepreneurial activities, from taking initiatives to apply for collaborative research grants, and so on. Lack of self-confidence makes her a better listener, but also less visible.

Another reason is the "good girl" syndrome. Women still have the instinct to set aside their own main interests in order to please and be loyal to the tasks other colleagues or a leading professor of their group hands out to them. To succeed in science, it is mandatory to focus on one's own main research interests and pursue them even if it leads to fights over money and power. This may include refusing to take on duties, which serve to defocus and distract the attention more than to increase visibility. Part of being a good girl is unfortunately to say yes to these tasks instead of no. We see the importance of doing general service to our research community more often than do our male colleagues. This inevitably leads to drawbacks and less visibility compared with our male colleagues, and to less time for research and research communication.

It also appears that women who avail themselves for positions in the power structure very often meet a negative response from their colleagues in doing so. It is not "allowed" for a woman to be aggressive, and a certain level of aggressiveness is often necessary to enter some of the power structures.

What is the situation in Norway?

In Norway we have a 40% rule applying to boards and committees of universities and public institutions regardless of field or level. This means that on any board of more than four persons, there should be at least 40% of both men and women. While this sounds good, there is a hitch. It works well in departments having at least 40% women colleagues, but it doubles the administrative burden of female colleagues in departments where only 20% or fewer are women. Typically these boards and committees handle teaching and faculty matters that, while being important within the faculty, are less beneficial for participation in the physics power structure. The main result is an administrative overload, which hampers the publication rate of women and the chance to benefit from the power structure networks.

There is no local leadership to protect women faculty from too many administrative duties; it is up to each person to refuse to take on more than they can handle. The result is that often, while several (not all) male colleagues can go without administrative duties for years, the female faculty member finds herself continuously on one or two committees/boards, and being blamed for having a slow publication rate. In turn, this severely hampers her access to the international power structure in physics. The Norwegian physics community is very small and thus not perceived as possessing important power positions.

Can this situation be remedied?

I think a number of things can be done. A systematic approach to nomination processes for invited talks might be one mechanism. Providing funds to meeting organizers for inviting female speakers is another incitement.

Also important is increasing the awareness among mentors of potential differences in their promotion of young female and male researchers. Female as well as male mentors would benefit from increased awareness.

Taking the mystery out of the power structures and networks is an important first step to gaining access through channels other than that of mentorship (from which most women researchers are probably left out).

Courses are needed, not only on leadership such as those that have been organized by Norwegian Universities for several years, but also on increasing awareness about how entrepreneurial activities and taking risks in launching new projects are important to women researchers. An awareness and knowledge on how to do this will also increase the self-confidence necessary to take initiatives.

Women in Physics in Pakistan

Asghari Maqsood[1], Fatima Hasnain[2], and Khalid Rashid[3]

[1]Quaid-I-Azam University; [2]APWA Government College of Women; [3]Quaid-I-Azam University

Pakistan is a country of 150 million people, (99% Muslims and 1% Christians). Because of a higher mortality rate for baby girls, the female population (49.5%) is slightly less than the male population. In the present prevailing social values, baby boys being more desirable than baby girls, the boys receive better care than the girls. The official literacy rate is given as 40%, but because of uncertain statistics, this figure is to be accepted with caution. The literacy rate for women is much less, perhaps about 10%. The Gross National Product in 2000 was US $56 billion, with a per capita income of US $429. Of the total expenditure of US $9.66 billion in 2001, 25% was spent on defense, 53% on debt servicing, 17% on development, 1.5 % on health, and about 1% on education and training.

EDUCATIONAL STRUCTURE

Education in Pakistan follows the structure introduced by the British as colonial masters around the year 1860. Primary schools go from class I to V, middle schools from class VI to VIII, high schools from class IX to X, and higher secondary schools and colleges from class XI to XII. After completion of class XII, students take the board's higher secondary school examination. The successful candidates with good grades usually join a professional college in engineering and medicine. The women candidates predominantly prefer medicine.

In colleges one continues studies for another two years in at least three subjects with equal weights to obtain a B.Sc. degree. Two further years of study at any of the state universities or postgraduate colleges leads to an M.Sc. degree in the subject chosen. Two more years of study at the state universities will lead to an M.Phil. degree. M.Phil. consists of one year of course work and one year of research. With this degree in hand the candidate can register for Ph.D. studies. In any one of the state universities, earning a Ph.D. usually takes three to five years. In some universities an M.Sc. degree is enough to register for a Ph.D. Some state universities also offer a B.Sc. Honors degree, which is somewhere between a regular B.Sc. and an M.Sc.

Education in all state schools and colleges is segregated with respect to gender (with very few exceptions at private institutions). Generally the laboratories are poorly equipped; facilities for research, especially in experimental physics, are extremely limited; and libraries have very few volumes. Teaching is mostly restricted to the prescribed syllabus, and rote learning from the notes of the teachers is the common mode of acquiring knowledge in schools and colleges. Concepts and problem-solving skills are not adequately stressed.

In some schools in the countryside or remote areas, even classrooms and chairs are not available and pupils have to sit on the floor in the open in the freezing cold in winter and extreme heat in summer. The private schools are much better equipped, but the exorbitant tuition fee is beyond the reach of the common man.

The numbers of students and teachers according to gender in Pakistan are summarized in Table 1.

TABLE 1. Male and Female Students and Teachers in Pakistan.

Level	Students (thousands)			Teachers (thousands)		
	Male	Female	Ratio	Male	Female	Ratio
Primary (I–V)	11,721	8679	74%	236	138	58%
Middle (VI–VIII)	2762	1882	68%	46	50	108%
High school	1157	775	67%	144	80	55%
College	420	372	88%	17	10	16%
University	87	27	31%	4.7	1.2	25%

CP628, *Women in Physics: The IUPAP International Conference on Women in Physics,* edited by B. K. Hartline and D. Li

PHYSICS EDUCATION

Higher physics education starts after obtaining a Bachelor's degree with physics as one of the three subjects. There are 11 state universities, one of which is a women's university. As a rule, there is no segregation by gender at the M.Sc. and Ph.D. levels. However, men's and women's postgraduate colleges are usually separate institutions.

The premier institutions are Quaid-I-Azam University in Islamabad in the north of Pakistan, Punjab University in Lahore in the middle, and Karachi University in the south. At Quaid-I-Azam University, the total number of M.Sc.'s produced from 1967 to 2001 was 531, of which 91 were women (20%). The number of M.Phil. was 364, of which 48 were women (15%). And the number of Ph.D.'s was 69, of which 13 were women, a ratio of 23%.

The physics faculty of Quaid-I-Azam University numbers about 16, of which three are women, one of whom is the Chairperson of the Physics Department. There are two women on the physics faculty of Karachi University, one on the physics faculty of Punjab University, and practically none on the physics faculties of other state universities.

Research at universities is partially supported by funding agencies in Pakistan such as the University Grants Commission, Pakistan Science Foundation, and Pakistan Atomic Energy Commission, and from outside Pakistan by the International Abdus Salam Centre of Theoretical Physics, and World Laboratories of the Third World Academy of Sciences. DAAD (German Academic Exchange Service) and the Alexander von Humboldt Foundation donate equipment and support for conferences and workshops to faculty who have had academic training in German universities under their respective programs.

The top research and development institutions, in addition to the universities where physics is one of the main research activities, are the Pakistan Institute of Science and Technology in Nilore near Islamabad, Optics Laboratories in Nilore, the National Institute of Silicon Technology in Islamabad, the National Physics Centre in Islamabad, and the Space and Upper Atmosphere Research Corporation in Karachi.

PROBLEM AREAS

Generally there are many problems and shortcomings in Pakistan in physics teaching and research that have their origin in social and cultural values and limited resources.

The medium of instruction in higher education is English. This is a problem very specific to Pakistan, because the students have to first learn to fully grasp the English language before being able to understand the contents of the lectures. At times the ideas being taught in English are too abstract because they do not connect to their vocabulary of everyday experience. Comprehensive textbooks in the official language, Urdu, do not exist, as is the case with the Chinese, Japanese, and Korean languages. To date we in Pakistan have not been able to find a satisfactory solution to this language problem that we have inherited from the British colonial structure.

Most teachers and students cannot afford to buy textbooks printed in the West. Laboratories at many universities are hopelessly underequipped. Most of the physics experiments at the M.Sc. level are in classical electronics.

Many a bright girl student is lost to sciences because of early marriage, subsequent family responsibilities, and lack of support from the family elders and husband to continue higher studies or professional occupation. Parents prefer their daughters to adopt either medicine or teaching professions because these have come to be accepted by the society as relevant women professions and offer part-time job opportunities. Many women today are attracted to business administration and information technology because these bring better rewards.

Women students of physics here are normally hardworking, but they do not find the environment of interaction with colleagues and are not allowed to work long hours in the laboratories. Parents want their daughters to be home as soon as classes end. After completing their degrees, very few women are allowed to practice their profession. This attitude, however, is changing rapidly because of the economic necessity of providing for the family. A single salary is usually not enough to survive on.

The challenge in Pakistan is to bring the excitement of physics to girls at an early stage and then to encourage them to become physicists and contribute their share to this wonderful endeavor of the human mind. However, here women in Pakistan need much support in substance and spirit from their colleagues in the richer Western countries.

ACKNOWLEDGMENTS

The authors wish to thank the IUPAP and APS for the financial support that made our participation to this conference on Women in Physics possible.

Women in Science in Poland

Izabela Sosnowska

Warsaw University

INTRODUCTION

Every scientist, man or woman, faces fundamental choices. For a woman, however, the most fundamental choice of all is whether to pursue her professional career or to take care of her family. There is no universal answer to this question, and every woman in academia must cope with the problem on her own.

On the surface it seems to be one of those questions that are well known and thoroughly discussed. Several articles have been written on how to reconcile family obligations with intense professional work. But typically these articles pose the wrong question: "How to help a woman make her decision?" instead of "How to change working patterns to help reconcile her conflicting desires?"

Many well-meaning persons have intensified this conflict by urging women to do both things the best they can. Men delight in praising "their own" females—wives and daughters—and no doubt they praise them to make them happy. Their advice is: "You are exceptional, you can do both, even if others cannot." These men wish to keep their wives and daughters at home, but they also want to be proud of them and make them believe it is not a sacrifice to devote oneself to the family. They assure their promising females that they rise head and shoulders above other competitors and force their women to cope heroically with a double load of duties.

The same men forget their generous and supportive attitude, however, if they confront a female stranger in their workplace. An intelligent and well-educated woman is an immediate threat, not only to their egos, but first and foremost to the special status of the domestic geniuses who are presumably an absolute exception to the rule that a woman cannot have her career and her family.

A sensitive man can overcome his instincts, if he tries, of course. So a new woman on the job is accepted sooner or later, in fact as soon as her male colleagues have had enough time to go through their period of cognitive dissonance. They eventually conclude that a female stranger can also be talented and intelligent if she is a part of their team, if she works hard and tries to be helpful, and, most importantly, if she looks like a man and acts like a man. But there is one rule that they will never break. She must absolutely never show any sign of aspiration to become the leader of the team. It is one thing for them to tolerate a woman that works as hard as a scrubbing woman—even if inadvertently she tars the picture of the female geniuses left idle at home. It is quite another matter, however, to be subjected to the decisions of a female that recently was no more than a stranger.

I am not the first one to draw attention to this problem, but we have a long way to go before we have fully grasped it and agreed how to solve it. I only want to emphasise one point. What most men expect in such circumstances is irrational and impossible to accept by an intelligent woman. They want her to subject herself to them in social and interpersonal relations before they consent to consider her a potentially acceptable boss. In other words, she must be inferior in order to be superior. There is no logic in such a wish.

These universal sociopsychological conditions are aggravated by some local circumstances. The real position of women scientists is best described in statistical data.

STATISTICS FOR POLAND

Women account today for a little more than 50% of Poland's population of 38.7 million (1998 data). This proportion is not reflected, however, in women's participation in scientific activities and research work and access to higher positions in education, research, and political institutions. Women in Poland, in the legal sense, have had equal educational and political rights since 1918. Throughout, even if they completed their university education and occupied progressively more highly responsible positions, their earnings were still lower than those of men.

CP628, *Women in Physics: The IUPAP International Conference on Women in Physics,* edited by B. K. Hartline and D. Li

Today one may say that women in Poland are free to decide about their education, even scientific or political, but this does not mean that their opportunities for professional, scientific, and political careers or their access to higher positions are equal yet. At 248 Polish universities and graduate schools of various kinds, there were 1,421,277 students in 1999, almost 60% (812,324) of whom were women. Statistics concerning postgraduate participation in scientific activities and political life indicate, however, that after university graduation this proportion changes. This is confirmed by the employment structures of Polish universities, institutes of the Polish Academy of Sciences, graduate schools, and research and development institutions. Statistics show clearly a considerable inequity in the scientific and political careers of men and women.

The percentage of women occupying full professor positions and the number of Ph.D. degrees issued in the same time frame confirm that the percentage of women decreases significantly at higher scientific posts (http://www.kbn.gov.pl/en/women_science/index.html).

As in other countries, activity in sciences, particularly physics and mathematics, is strongly dominated by men in Poland. Women do not take up work in physics or engineering proportionately to their talents and abilities. Furthermore, those who have joined the scientific communities give up further professional development for social reasons.

At the Faculty of Physics of Warsaw University, women account for about 35% (average for the last 6 years) of the total number of students (about 900). Women constitute 41% of total number (120) of Ph.D. students. But the number of women at higher levels of scientific careers is much smaller. Only 5 women are among 48 full physics professors, 7 among 34 associate professors, and 27 among 120 assistant professors, senior lecturers, and lecturers. This is a much lower percentage than the average for the whole university. The Physics Teachers College at the Physics Faculty of Warsaw University presents an exception—women account for about 60% of total number of students

The Polish government instituted the National Program of Activities for Women in April 1997 to support participation of women in scientific and political life. In the previous parliament women accounted for only 13%, and in the present one (September 2001 election) their number amounts to about 20%.

THE FUTURE

I am not going to offer any wide-ranging solutions. But I want to suggest one remedy that can help us find a more reliable settlement in the future. Women who already possess high academic status should oppose without hesitation every nomination of a poorly qualified woman to an important position. We should get rid of the "token female superiors" that are put in high places to create the belief that discrimination has ended.

An incompetent female boss is an attractive compromise for many quarreling parties. Those who distrust women in general are satisfied because one more woman has been discredited as an incompetent boss. Those who claim that discrimination has ended are satisfied too, because they have a clear example of a female boss, good or bad, but with high professional status. Many women are also happy because such an example raises new hopes. If they accept a poorly qualified woman as a boss—assume the aspiring females—they will be even happier to accept a better-qualified woman in such a place.

But all this is a complete illusion. Make-believe accomplishment does more harm than good. Nobody needs female bosses that are tolerated but not respected. We should not be pacified by examples of females that are endowed with high administrative powers, if the real outcome of their rule is a widespread belief that women are unfit to serve in position of high responsibility.

The bold women who have made a successful academic career should take their achievement seriously. They should thwart any attempt to discredit its real value. If we have satisfied high standards of academic excellence in the highly competitive world, it is an insult to our achievement if other females are offered comparable benefits without being put to similar tests. Let us oppose defamation even if it is done with good intentions, ostensibly to equalize the balance between men and women. We should also oppose the stereotype that women are inept professionals but effective operators—scheming, ruthless, and ambitious. Maybe some are, but most of us are not.

So if you read one day that Maria Skłodowska-Curie was a brilliant scientist, but nevertheless an unscrupulously manipulating woman that seduced a "young academic genius," think twice before you believe it. Do not look for sinister causes behind professional achievement of every bold and independent woman with or without family. Be satisfied with the thought that some of us deserve what we get. Help others to get the same treatment.

A Physics Career for Women in Portugal

C. Providência[1], M.M.R. Costa[1], and A.M. Eiró[2]

[1]University of Coimbra; [2]University of Lisbon

A recent study shows that among European countries Portugal has the highest percentage of women researchers in higher education: 48% in natural sciences and 29% in engineering and technology (Table 1). As in many other Latin countries, and in contrast to the northern European countries, Japan, or the United States, women are present in physics departments of the public universities in Portugal in quite a large percentage, at about 28% (Table 2).

WHAT MAY EXPLAIN THIS BEHAVIOR?

There is in our country a very open-minded attitude toward the subjects chosen by young boys and girls in schools. At that level both genders are equally represented in all classes—sciences, arts, or others. Physics courses, however, do not attract many students, mainly because it is a difficult subject and does not lead easily to well paid jobs. Most physics majors get a job as a teacher in secondary school. Some follow a university career, usually the brightest students, and only a very few are admitted to industry. There is almost no research being done in industry; engineers compete with physicists for jobs and are very often preferred by employers.

Although among students we have equal gender representation, in the choice of professional careers that equilibrium is broken. For example, 74% of the high school teachers of physics and chemistry are women.

Women in Portugal can achieve as high a level as men in any professional structure when there are no family restrictions, and a physics career is no exception. Nevertheless, this is far from being the most common situation.

TABLE 1: Female Researchers in Higher Education in Europe.

Country	Natural Sciences (%)	Engineering and Technology (%)
Portugal	48	29
Ireland	44	25
U.K.	31	14
Italy	31	13
Finland	29	19
Sweden	29	18
France	29	17
Denmark	23	13
Austria	18	9
Germany	14	9
Belgium	11	2
The Netherlands	8	6

Source: European Commission, Eurostat. Reprinted with permission from Science *295*, 41 (2002). Copyright 2002 American Association for the Advancement of Science.

CP628, *Women in Physics: The IUPAP International Conference on Women in Physics,* edited by B. K. Hartline and D. Li
© 2002 American Institute of Physics 0-7354-0074-1/02/$19.00

TABLE 2. Women in Physics Departments in All Public Universities in Portugal.

University	Women	Total	%	Full Professor			Associate Professor			Assistant Professor		
				Women	Total	%	Women	Total	%	Women	Total	%
U Algarve	6	21	29	1	1	1	1	4	25	3	11	27
U Aveiro	13	54	24	4	7	57	3	12	25	3	22	14
UBI	6	34	18	0	0	-	0	5	0	1	7	14
U Coimbra	24	71	34	4	9	44	2	10	20	12	34	35
U Évora	6	30	20	0	3	0	1	8	13	2	8	25
U Lisboa	21	62	34	2	10	20	6	12	50	12	40	30
U Minho	17	56	30	1	5	20	4	7	33	4	21	19
UNL	10	34	29	1	3	33	2	6	33	3	12	25
U Porto	11	35	31	1	8	13	3	8	38	7	19	37
UTAD	4	24	17	0	0	-	1	3	33	0	3	0
UTL	17	66	26	0	9	0	3	11	27	14	46	30
TOTAL	135	487	28	14	55	26	26	86	30	61	223	27

Salaries are low and couples cannot live on a single salary, meaning that all women have to work outside the home. However, society still considers that the career of the husband should lead the options of the couple. Teaching jobs, being generally more flexible than many others, are preferred by many women. There are organized structures to look after children, and a general attitude for the extended family to collaborate in the bringing up of children. This was particularly true one generation ago when older women did not have professional careers and it was relatively inexpensive to pay for a full-time person (always a woman) to look after children.

Some years ago, the system in public universities was not extremely competitive, and although career progress was slower for a woman, promotion was almost always achieved—that explains the numbers of Table 2. The situation has changed completely. Promotion now implies an international career, and it is harder and harder to accommodate this with family life. We anticipate that the situation that made Portugal an example for the representation of women in science will slightly change in the next decade.

The compatibility of a professional career with a family is an unsolved problem in our society, with no perfect solution. This problem can be partially solved with flexible working hours, flexible working places, the existence of support structures for children, support from employers, and, especially in academies, being released from administrative duties and not overburdened with teaching.

Cultural, Societal, and Professional Features of a Woman's Career in Physics in Romania

Anca Visinescu[1], Sanda Adam[1], Violeta Georgescu[2], and Agneta Balint[3]

[1]National Institute of Physics and Nuclear Engineering; [2]Alexandru Ioan Cuza University; [3]West University

The beginning of a school of physics research in Romania can be traced back about a century. The "first wave" of women physicists doing systematic research in various branches of physics in our country started at the beginning of the 1960s. Historically, the number of women full professors in physics has been small and is still small.

Far fewer girls than boys finishing high school with a background in mathematics and natural sciences were potential candidates for a career in physics in the 1950s and the 1960s. Although there has not been a dedicated study, our immediate experience places the girls/boys ratio in the 1960s between 1:3 and 1:5. This was a result of the European tradition that assumed that an art-based profession is much more appropriate for a young girl than one based on science.

The post-war "socialist" rule resulted in a dramatic inversion of this ratio toward the end of the1970s and 1980s: the number of girls getting a high school diploma with a basis in mathematics and natural sciences surpassed that of boys. As a result of this trend, which reflected the lack of interest of boys toward "knowledge for the sake of knowledge," the number of women graduating in physics significantly increased. However, the number of women getting a Ph.D. degree and working in research institutes and in university physics departments was by far lower (about 20%).

A possible explanation for this apparent paradox is to be found in the school system developed during the "golden epoch" (Ceausescu's period), when the free spirit stimulating freedom of thought was not considered as something fundamental. In the purely scholastic vein, knowledge was regarded as coming from the communist party leaders and the founding fathers of the "scientific socialism," rather than the result of unrestricted, innovative minds. In this spirit, it was the quantity of knowledge, rather than its quality and the process of its discovery, that was offered to youngsters, both girls and boys.

Moreover, the pyramidal structure of society resulted in the development of impermeable "castes," with little possibility of penetration by people from abroad. Thus, the fundamental trend of the so-called "socialist society" was that of involution, such that it came as no surprise that a popular uprising took place, which changed the fundamental rules of the political game toward democracy.

The last dozen years have been a period of transition. The above-mentioned ratio established during the 1980s has remained about the same (yet in 2001 a roughly equal number of men and women obtained the first-level degree). However, for the most gifted girls and boys finishing high school, getting a graduate degree in physics in a Romanian university has ceased, to a large extent, to be as attractive as before. Most of the best young people, who have been identified during high school through the international contests of mathematics and physics (named "Olympiads"), are starting their higher education in universities in the United States, where they find conditions excellent for their future development.

Among those finishing their studies in Romania, the dream of coming to the United States and Western European countries is realized upon completing their Ph.D. degree. However, only a small ratio of these students returns to Romania. Many of those returning home find it more profitable to change their field of activity—many of the most prosperous businessmen in Romania today are former physicists.

This lesson learned in the western countries will, undoubtedly, bear fruit over time. For the time being, however, the profession of physicist in Romania is a difficult one, with little possibility of getting financed. Figure 1 shows the decreasing interest for studying physics. The low wages offered physicists in research and education is the main reason for the decreasing number of students in physics.

CP628, *Women in Physics: The IUPAP International Conference on Women in Physics,* edited by B. K. Hartline and D. Li
© 2002 American Institute of Physics 0-7354-0074-1/02/$19.00

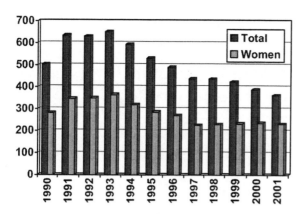

Figure 1. Total number of students in physics at the University of Iassy.

Twelve years cannot fundamentally change the situation of young physicists in Romania, male or female. The difficult conditions of life as a physicist are overwhelming in this country, and the experience of the women is even more difficult due to their supplementary motherhood obligations. There have not been explicit discriminatory actions against the access of young women to higher scientific and academic positions in physics, but the worsening of the status of physics on a worldwide scale has added more burdens to their shoulders.

In the last year, only 19% of the Ph.D.'s in physics were awarded to women, although about 49% of the graduate students in physics were women.

In a study of the male-female asymmetry in physics, the sociological aspects involved in the process of becoming a physicist are more important than the statistical details. An important argument for the small number of women in physics rests in the most familiar differences between men and women: the differences between social and biological forces. The education-governed actions that form a child into a physicist could include parental interaction, toy selection, and childhood and adolescent interactions with peers.

The role of female models to encourage young women to pursue their interests in physics is a very important one. The usual images promoted by the mass media of women as supermodels or seductresses serve to lure adolescent females from a path of science. There is a strong need for the promotion of an opinion among women themselves concerning their careers in science.

Perhaps the best we can hope for is to be stubborn enough and healthy enough to act toward the creation of a favorable background for wiser decisions of the entire society.

Russian Women Physicists in the Transitional Period

Galina Ya. Merkulova[1], Nelly I. Didenko[2], Eduard A. Tropp[3],
and Olga S. Vorontsova[4]

[1]St. Petersburg State Technical University; [2]St. Petersburg Research Center of Russian Academy of Sciences (RAS);
[3]Ioffe Institute, Russian Academy of Sciences; [4]Sarov, Nizhny Novgorod Region

In the present-day transitional period in Russia, professional and social adaptation of all scientists, including women physicists, is ongoing. The first stage of the transitional period, the early 1990s, was marked by migration of physicists to other fields of activity, long-term business-trips abroad, and even work abroad through emigration. The second stage, from the mid-1990s, was characterized by cooperation with the world community of physicists.

Earlier in the Soviet Union it was quite prestigious to be a scientist, and the engineering sciences attracted particular attention. Russian schools of thought in physics and mathematics are traditionally strong. The interests of the Soviet military-industrial complex were an additional stimulus to study the natural sciences, including physics. Furthermore, equal rights between men and women under our constitution actually existed for young men and women in choosing a profession, and women did not hesitate to choose engineering sciences over the humanities. Unfortunately, women did not have, and still do not have, equal opportunities in their professional career growth.

To analyze the current state of women in physics, one should distinguish the work of women physicists in Russian Academy of Sciences (RAS) institutes and universities from their work in applied research institutions. In the early 1990s, many young women physicists left academic research laboratories following a dramatic cut in state financing. In the mid-1990s, most active women physicists from the RAS were intensively involved in scientific projects supported by Western and Russian scientific foundations, whereas in applied science research, entire laboratories closed and scientific groups disappeared, especially at ex-military institutions. As a result, many women physicists with high qualifications went into business organizations such as banks and trade companies.

The percentage of women in physics in St. Petersburg State University is 27% with master's degrees; only 7% are full professors. In the St. Petersburg RAS institutes the percentage of women in physics varies from 24% (without scientific degree) to 13% as Doctors of Science.

The current state of Russian women in physics can be visualized as a pyramid: at the base, or at the beginning of a career, you find more women: about one-third of the total physicists at that level. The higher the position, the lower the number of women. According to the State Financial Academy in 1995, fewer than 5% of women occupy leading positions in the natural sciences, and the number of women among the Doctors of Sciences is fewer than 13%. In Russia there is currently not a single woman "academician" (elected member of the RAS) in physics, whereas there are dozens of men academicians.

A problem for Russian science in general, and physics in particular, is the aging of scientists. A real decline in the age group of 25- to 50-year-olds is caused by transitional problems that resulted in the migration of many scientists to business or foreign countries after receiving their Ph.D.'s—the male migration being more considerable. Data of the Russian Foundation for Basic Research (RFBR) show a "hole" in the age distribution for men participating in RFBR projects, but for women the curve is smooth (see Figure 1). Women remain in physics—so we can consider them our reserve for physics now.

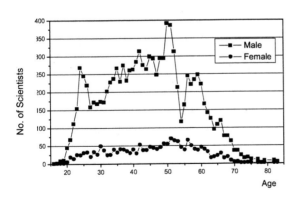

Figure 1. RFBR scientists in physics and astronomy in 1997. (Data from Vladimir A. Minin and Alexandr N. Libkind, private communication.)

CP628, *Women in Physics: The IUPAP International Conference on Women in Physics,* edited by B. K. Hartline and D. Li

BARRIERS TO A SUCCESSFUL CAREER FOR A WOMAN PHYSICIST

Women often play secondary roles because of traditional Russian stereotypes. Historically, a woman has to take care of the family and household. Huge social changes have actually led to inequality between men and women. In 2000 David Wolfe wrote in the *CSWP Gazette*, "Russian women-scientists are fighting problems of discrimination, salaries, job placement, childcare—at levels against which we [in the USA] struggled at least fifteen years ago."

Low self-esteem is also a major obstacle to professional success for Russian women. Russian women physicists are inclined to underestimate themselves more than their Western colleagues. They accept more easily when they are not paid for their work or are paid too little. In physics a man sometimes has the title of project chief, while a woman is the person who does the real work. This situation is typical, especially in applied research institutions.

Going up a career scale, women encounter the so-called "glass ceiling," which is difficult to overcome: public opinion is reluctant to see a woman as a leader in natural sciences. In any society, women have to put forth greater efforts than men to demonstrate their professional capabilities. In our society, in which science is overburdened with bureaucracy, these efforts are really tremendous. It appears that bureaucracies are conservative structures that function according to a certain algorithm that was formed long before women realized their rights.

SUCCESSFUL WOMEN PHYSICISTS IN RUSSIA

Many women physicists, mainly from older generations, have made great contributions to the world of science. Many of these women are inventors and patent holders. For example:

- Maria Chaika, Professor and Laboratory Chief, St. Petersburg State University; Honored Science Worker, State Prize winner. More than 150 articles, 2 monographs; registered discovery in optic orientation in gases.
- Nelly Esepkina, Professor and Chief of Optoelectronics in Computer Science, St. Petersburg State Technical University. Over 150 articles; deputy chief designer of world's largest reflector telescope, the RATAN 600.
- Iya P. Ipatova, Professor and Principal Researcher of Ioffe Research Institute of RAS. More than 150 articles, including the brilliant one devoted to nonequilibrium decomposition in semiconductors.
- Galina Kandaurova, Professor, Ural State University. Principal researcher of the physics of metals, magnetism, the theoretical and experimental study of different-domain magnet structures.
- Ekaterina Feoktistova, Doctor of Sciences, a three-time State Prize winner. She passed away in 1987 having devoted her entire life to work in institutes of the Ministry for Atomic Energy of Russia. She studied properties of explosives and their use in Soviet nuclear weapons, and much more.
- Irina Glushneva, Professor, Moscow State University. More than 130 articles on star photometry; author of photometric star catalogs.

HOW TO ATTRACT WOMEN TO PHYSICS?

Recently many young Russian men left physics for business, leaving spaces for women to occupy. This is reflected in the increase in women postgraduates in physics—from 7% to 20% over the last five years (Ioffe Research Institute, RAS). The real problem is to retain these women in the field. The most ambitious of them go abroad for training and work, and quite often marry and stay there for a better life. We need to revive an interest in physics with higher wages and more cooperation with the international scientific community. We need to raise the prestige of the work of women in physics and to prime women scientists for success both mentally and financially.

The social activity of women physicists is low, because they are engaged primarily in survival in new economic conditions. In 1999 the St. Petersburg Women Association in Science (SPWAS), a nongovernmental organization, was established. In 2000 SPWAS held its first international conference, "Women in Fundamental Science," where the results and prospects of interdisciplinary research were considered. This conference received financial support from RFBR and academician Zhorès Alferov, the Vice-President of RAS and a Nobel Prize winner.

ACKNOWLEDGMENTS

Many thanks to the following for the data they provided us: Vladimir A. Minin, head of RFBR's Department of Methods and Analysis; Alexandr N. Libkind, Senior Specialist of RFBR; Vladimir N. Troyan, Vice-Rector of St. Petersburg State University; and Yuri N. Fokichev, head of Personnel for RAS St. Petersburg Research Center.

Education and Employment for Women in Physics in Slovakia

Marcela Morvová[1], Eva Majková[2], and Zuzana Dubničková[1]

[1]Comenius University; [2]Slovak Academy of Sciences

The participation of women in Slovakia in technical subjects, including physics, has a relatively long tradition. After World War II, many women in Slovakia began to build their professional careers, particularly in physics. Primary and secondary schools were all coeducational, and there were no obstacles for girls in mathematics, natural sciences, and technical education.

In spite of relatively good educational possibilities, the number of women participating in research is not as high as in several European (especially northern) countries. The main reason is the lack of the necessary infrastructure to allow women to balance a family and career. Another big problem is that the low salaries at universities and research organizations hinder women from hiring help for the household, forcing some women to have two jobs.

Research in physics is done in the Slovak Academy of Sciences (SAV), in governmental research institutes, and in institutes with state participation, such as the Slovak Institute of Metrology, the Nuclear Regulatory Authority of Slovakia, VÚZ (the Research Institute of Welding), and VUJE (the Research Institute for Nuclear Energy). There are also several engineering organizations that perform applied research, but data for evaluating the participation of women are not available for these organizations.

The data for the evaluation of the participation of women in physics in Slovakia were collected from educational organizations and from SAV, and are presented in the remainder of this paper. Only persons with appropriate physics education were included in the statistics under "staff."

Comenius University: The Faculty of Mathematics, Physics, and Informatics of Comenius University provides university education in mathematics, physics, and computer science, as well as teacher training in subjects related to these branches of study. The faculty offers an M.Sc. degree (5 years) and a Ph.D. degree (additional 4 years) in astronomy and astrophysics, biophysics and molecular physics, electronics, plasma physics, solid-state physics, geophysics, nuclear and subnuclear physics, meteorology and climatology, optics and optoelectronics, theoretical and mathematical physics, applied physics, environmental physics, and biomedical physics. Teacher training forms a separate group of study, where students can choose one or a combination of two subjects. Of 190 staff members, 41 are women. In 2001, nine women were awarded a Ph.D. degree. The gender breakdown for M.Sc. degrees for the last five years is shown in Table 1.

TABLE 1. M.Sc. Degrees in Physics Disciplines at Comenius University, 1997–2001.

	1997	1998	1999	2000	2001
Women	35	20	23	40	28
Men	72	57	74	58	44

Slovak University of Technology in Bratislava: The Faculty of Electrical Engineering and Information Technology of the Slovak University of Technology in Bratislava offers a three-stage system: bachelor's program (4 years), master's program (1.5 years), and doctoral program (3 years internal, 5 years external). Bachelor's and master's degrees can be obtained in automation, electrical materials science, electronics, power engineering, and information technology. Of 214 staff, 42 are women. The women/men ratios for degrees awarded in 2001 were: 176/704 B.Sc., 36/144 M.Sc., and 8/61 Ph.D.

University of Žilina: The University of Žilina was established in 1953. Its basic engineering degree course (corresponds to M.Sc.) takes 5 years. A postgraduate program that leads to a doctoral degree takes 3–4 years. Some fields offer a bachelor's degree that takes three years. All of the university's faculties provide a supplementary course of pedagogical studies for students and graduates. The university has the following faculties: Operation and

CP628, *Women in Physics: The IUPAP International Conference on Women in Physics,* edited by B. K. Hartline and D. Li

Economics of Transport and Communications, Mechanical Engineering, Electrical Engineering, Civil Engineering, Management Science and Informatics, Natural Sciences, and Special Engineering.

The Department of Physics formally belongs to the Faculty of Electrical Engineering, but provides physics education for the engineering students of all faculties of the university. The department is divided into three divisions according to the areas of interest of its members: Acoustics, Optics, and General Physics. Of 44 department staff members, 12 are women.

Matej Bel University: Matej Bel University in Banská was founded in 1992 by the Pedagogical and Economics faculties. It now contains eight faculties. The Faculty of Natural Sciences is primarily known as a teachers' faculty, in that it mainly prepares teachers in general education subjects for grammar and secondary schools. The physical sciences are taught by the Department of Physics, which has a staff of 22 teachers (five women, 17 men) and is currently preparing two female Ph.D. students.

University of P.J. Šafárik: The Faculty of Natural Sciences of the University of P.J. Šafárik in Košice prepares teachers for grammar and secondary schools in mathematics, informatics, physics, chemistry, biology, and environmental science. Women comprise 22 of its 72 staff members and two of its 13 Ph.D. candidates.

Slovak Technical University in Košice: The Faculty of Electrical Engineering and Information Technology at Slovak Technical University in Košice was founded in 1969 and offers a three-stage system: a bachelor's program (3 years), a master's program (5 years), and a doctoral program (3 years). In 1997 the faculty was divided into a faculty in Košice and one in Prešov. The faculty remaining in Košice provides M.Sc. courses in computers and informatics, electronics and telecommunication systems, electro-energetics and power engineering, control engineering and automation, technologies in electronics and materials, electrical measurements, and industrial engineering. Of 93 staff, 34 are women. One of the faculty's seven Ph.D. candidates is a woman.

University of Prešov: The University of Prešov has five faculties: Arts; Humanities and Natural Sciences; Education; Orthodox Theology; and Greek-Catholic Theology. Because of the Department of Physics, the Faculty of Humanities and Natural Sciences also offers pedagogical education in physical sciences. Three of the 11 physics staff are women. The women/men ratios for M.Sc. degrees awarded in recent years have been 4/12 in 1997, 7/8 in 1999, and 8/10 in 2001.

University of Konstatin Filosoph: The University of Konstatin Filosoph in Nitra was established in 1959 as a pedagogical university. It prepares teachers for primary and secondary schools and specializes in material physics provided by the Department of Physics. The department has 17 staff members, of whom five are women. One of its five Ph.D. candidates is a woman.

University of Agriculture: Established in 1958, the Department of Physics is one of the oldest departments at the University of Agriculture in Nitra. Four of its eight staff are women, and the department has one Ph.D. candidate, a man.

Slovak Academy of Sciences: The Slovak Academy of Sciences is the largest research institute in Slovakia. It is divided into the physical, life, and social sciences, and has six physical institutes: astronomical, physical, geophysical, experimental physics, measurements science, and electrotechnical. The gender compositions of these institutes are shown in Table 2.

TABLE 2. Staff Compositions of Physical Sciences Institutes at the Slovak Academy of Sciences.

	Astronomical	Physical	Geophysical	Experimental Physics	Measurement Science	Electro-technical
Women	1	13	12	1	25	21
Men	16	62	14	15	56	91

National Totals: According to the data above, the percentage of women participating in research and as pedagogical staff at all universities in Slovakia is 24.1%, the percentage of female physics students is 24%, and the percentage of female physics doctoral students is 14.2%.

Women in Physics in Slovenia

Norma Susana Mankoč Borštnik[1], Maja Remškar[2], and Andreja Gomboc[1]

[1]*University of Ljubljana;* [2]*Jozef Stefan Institute*

GENERAL STATEMENT AND HISTORICAL VIEW

One of the slogans of socialist countries is "Equal Opportunities for Men and Women." Until 10 years ago, Slovenia was a part of a socialist country. The declaration of equal opportunities for men and women helped a bit, especially since families could not live on the incomes of men only. The state arranged kindergartens, which was helpful for working mothers.

However, as in all countries, the tradition that a man should be the leader at home and accordingly also in companies, and the fact that women spend on average more time with children than men, made the realization of equal opportunities rather weak. The transition of my country to capitalism did not make this situation better. Only a few women can be found in leading positions in politics or in companies.

The fact that elementary and middle schools in Slovenia, as in most European countries, collect more and more data in their programs that teachers do not really understand (even their colleagues at universities many times do not—new discoveries take time to be properly understood and pedagogically presented), and accordingly cannot explain to pupils in a way that pupils can follow and enjoy, forces pupils to memorize without understanding. Because girls are on average more willing to obey requirements than boys, girls are more successful in gathering good marks and so have more possibilities for university studies. For this reason, lawyers, physicians, and economists are mostly women.

In physics, teachers in the elementary and middle schools are mostly women, but not researchers and university teachers—they are mostly men. Self-confidence (men much more than women are stimulated by tradition to be leaders) could be a reason for that. Another reason is that research work takes time, and women often have to take care of children when they are young (usually much more than their partners), losing the competition for positions.

THE STATISTICS

We present some illustrative data, most averaged over the last 30 years:

- The relative number of women who finished bachelor's degrees in physics: 16%.
- The relative number of women who obtained master's degrees in physics: 15%.
- The relative number of women who obtained Ph.D.'s in physics: 16%.
- The University of Ljubljana, Faculty of Mathematics and Physics, has one female full professor with special status (7%), one female full professor (7%), no female irregular professor (out of 7), no female docent (out of 7), and four female assistants (31%).
- The research staff of our country's national research institute for physics, Jozef Stefan Institute, is 10% female, including Ph.D. students.
- The Ministry of Science of Slovenia instituted a national program for young researchers 15 years ago. Women have comprised 19% of the postgraduate students in physics there (30 of 160 students); to date, half of them have received their Ph.D. degrees. Four women, or 27% of those who obtained their Ph.D., have obtained permanent positions at Jozef Stefan Institute. By comparison, the corresponding number of men is 30, or 50% of those who have obtained their Ph.D. degree.

CP628, *Women in Physics: The IUPAP International Conference on Women in Physics,* edited by B. K. Hartline and D. Li
© 2002 American Institute of Physics 0-7354-0074-1/02/$19.00

WHAT CAN BE DONE TO IMPROVE THE SITUATION?

It is, of course, in the interest of any state to make a positive selection among candidates for positions, regardless of gender, and to stimulate everyone to learn as much as one can. Tradition is, however, a strong presence, and men are not willing to release their privileges, nor are women easily stopped from supporting and trusting their own men more than themselves. So changes go slowly.

A good educational system is very important. Everyone should be stimulated to learn in a way that she or he can understand in order to integrate the knowledge. However, education, too, is a slowly changing process, because teachers themselves cannot change quickly either, especially because those who need to stimulate change have to learn themselves how to make changes in the right direction.

It is important to have role models who demonstrate that things can go differently in a better way. One of us (SNMB) would suggest, for example, a foundation of an international faculty for physics and other natural sciences and technologies in Slovenia, and in other parts of the world, in which the best women would get half of positions and be able to demonstrate that women in science are as good as men are.

One of us (SNMB) has tried, as a member of the Board for General Education in Slovenia, to influence the way of teaching natural sciences and mathematics in middle and elementary schools so that the amount of knowledge that pupils retain would increase. (We have a very European way of teaching.) The influence was small, because a teaching system can only change very slowly and changes when teachers change, which takes time. The influence of IUPAP on how physics can be presented so that pupils can acquire knowledge could be very important.

South Africa: The Rainbow Nation, Women and Physics

J. Padayachee and E.C. Viljoen

National Accelerator Centre

"Apartheid." A word that ruled South Africa since 1961; a word that meant that there was limited opportunity, often none, for non-Whites. It officially ended on April 27, 1994, a momentous day in history that will go down as South Africa's first democratic election. More importantly, it will go down as the day South Africans took the first steps to restore the balance that had been upset by apartheid. South Africa is presently in the midst of great social restructuring, with South Africans being more open to change. This is the perfect time to highlight the problems of women, and of women in physics, because there is great hope for change.

South Africa is unlike most countries in Africa. It has a large developed-world infrastructure merged with all the problems typical of a developing-world country. High rates of poverty, illiteracy (the 1996 census showed that 36% over the age of 20 were illiterate, 55% of whom were women), unemployment, crime, HIV/AIDS, and related issues are South Africa's major social problems. South Africa needs the economic growth that comes from technological development to pave the road to the future. Since scientific thinking drives technological development, South Africa desperately needs scientifically literate people. However, in a country where the need to survive comes first, scientific literacy is not given much priority.

After the 1994 elections, affirmative action policies were implemented across the country, in all walks of life, to give preference to previously disadvantaged or underrepresented groups, including non-Whites and women. This has helped get women into careers where there were none before.

PHYSICS IN A SLUMP

Physics in South Africa is in a slump at the moment. This can be seen in schools, at the undergraduate level, and at the research level. Prior to 1994, 90% of the country's education funds went to 20% of the population (White). Blacks were largely excluded from an education in science and mathematics. Although 10 years of education is now compulsory and free for all South Africans, improving science and mathematics education is extremely difficult due to lack of infrastructure and teachers. Also, many underqualified teachers are teaching science. In 2000, 60% of South African schools had electricity, 64% had access to telephones, and 12% had access to computers. Many South African schools do not have access to books or laboratory equipment to perform the most basic of experiments.

South Africa has 11 official languages. While English is often the language of instruction, a Black language is the language of understanding.. The government has released a plan for increasing the number of teachers and improving the standard of science education (http://education.pwv.gov.za/), and all schools should have electricity and telephone access in three years.

At universities, students see an uncertain, low-paying career in physics research and opt for the safer, high-profile jobs offered by the engineering, IT, and finance sectors. Also, fewer bursaries are offered to study physics. This has generally resulted in a steep decline in the numbers of students choosing physics at university. There are, however, certain physics departments (e.g., at the University of Natal-Pietermaritzburg) that have seen their number of students grow, partly because of program to help unprepared students.

At the research level, government funding has not increased over the past few years. Many companies that employ physicists have undergone some form of restructuring, forcing many who lost jobs to leave the country.

WOMEN, SOCIETY, AND PHYSICS

The average South African woman faces many hardships in her daily life, with violence against women being a dominant problem. The government has done a great deal to empower women, as can be seen by the number of women now in high-profile positions, but it is difficult to improve the quality of life of all women.

Women who choose a career in physics have problems of their own. They are discouraged from studying physics from an early age by parents and teachers; they do not fit into the male-dominated research environment; they have to work harder for promotions, funding, and respect; and they find it difficult to be taken seriously by male colleagues. Women in the workplace are often treated as outsiders and many feel unwelcome. Most White men in South Africa have no idea what it feels like to be an outsider in a workplace—how it eats at one's confidence and drives one to look for other jobs or turn down potential employment.

On the lighter side, female physicists seem to have greater success than their male colleagues with getting technicians to help them out!

Physics: Out of the Lab!

Conventionally, choosing physics means choosing academia and research. It must be emphasized to society that choosing physics means being trained to solve problems. A result of the abstract nature of physics is that students learn to handle abstract problems and are then better equipped to tackle problems of any nature from any sector. While most South African physicists choose academia and research, there is a small percentage that have chosen to go into sectors ranging from business to industry to finance.

As was concluded at the 2000 IUPAP Conference on Physics and Industrial Development (COPID 2000), for physics to thrive in the developing world there has to be greater alignment between physics, industry and the developmental needs of the country. The contributions to innovation made by physicists working in industry, not only academia and research, must be used as selling points to prospective physics students, raising the profile of physics and , hopefully attracting more women to the field.

Physics must not be sold as a "genius" subject. Children should not be told (as they often are!) that they are not smart enough to do physics. Girls must be encouraged to explore the beauty of nature from the physical sciences' point of view. The encouragement for girls to pursue physics should come from those women who have already done so. This means that female physicists must be made more visible to society by, for example, visiting schools to expose young girls to physics. These visits must be coupled with a science-awareness program for their parents and teachers to encourage girls to choose science.

Lending a Helping Hand

Much is being done to improve the status and working conditions of women in science. There is strong support from the National Research Foundation (NRF), the South African science funding entity, encouraging established researchers to take on female students. There is also pressure from top levels at universities to employ women, but sadly there has been little response to advertisements for these posts.

The "Women-in-Research" program (http://www.nrf.ac.za/wir/), established by the NRF seeks to stimulate discussion and address key issues that affect women researchers. Of particular concern is the serious underrepresentation of women among senior researchers, heads of departments, senior management, and those who access resources from funding agencies and organizations. The program aims to support women, especially non-Whites, to develop and strengthen their research skills and to increase the number of women in postgraduate studies, academia, research, and leadership positions at tertiary and research institutions.

The Association of South African Women in Science and Engineering has carried out many awareness activities in order to raise the profile of women scientists in South Africa. They often organize trips for young girls to promote science as a career. They also provide a bursary for honors study in science or engineering.

The South African Institute of Physics (SAIP) is the only professional body for physicists in the country. Unfortunately, the SAIP has been dominated for most of its existence by White, Afrikaans-speaking men, thereby alienating many (even male) physicists. Fortunately, the composition of the SAIP has been changing over the past few years and the institute is presently involved in a transformation process to bring it in-line with the current needs of physics and physicists in South Africa. There are now three women on the council of 10. 2001 also saw the first woman elected to the presidency of the SAIP in its 46-year history.

At the annual Conference of the SAIP in 2001, the Vice-Chancellor of the University of Natal pointed out that the future of physics lies in the hands of the physicists. Perhaps we, the women in physics in South Africa should band together and take the initiative to improve our future and increase our numbers.

ACKNOWLEDGMENTS

The authors wish to thank Catherine Cress, Patricia Whitelock, and Kevin Meyer for their contributions.

Women in Physics in Spain

C. Carreras[1], M. Chevalier[2], E. Crespo[1], M. García[3], M.T. López Carbonell[4], P. López Sancho[3], P. Mejías[2], R. de la Viesca[3], R. Villarroel[5], and M.J. Yzuel[6]

[1]*Universidad Nacional de Educación a Distancia;* [2]*Universidad Complutense de Madrid;*
[3]*Consejo Superior de Investigaciones Científicas;* [4]*Iberdrola;* [5]*Consejo de Seguridad Nuclear;*
[6]*Universidad Autónoma de Barcelona*

Historically, physics has received little attention in Spain for sociological and religious reasons. During the 16th and 17th centuries, Spanish universities were closed to outside influences in order to prevent heresy while Spanish kings were fighting reform in Europe. The Spanish Physical and Chemical Society was founded in 1903, and in 1907 the first physics laboratory was created by the *Junta de Ampliación de Estudios* (JAE, Institute for Advanced Studies). From 1936 to 1939 scientific activity was cut by the civil war and many scientists left Spain for political reasons.

The first contact with physics in school is at the age of 13. Physics is taught together with chemistry, and the teachers are usually chemists. The first-level degree in physics is the *Licenciado en Físicas*. It is a degree of four or five years; 19 universities offer this degree. About 25%–30% of first-level degree recipients are female. The highest-level degree in physics is the Ph.D., or *Doctor*. Although the ratio of female physics professors is about 25%, the highest position, full professor, is only reached by 3% of women.

TABLE 1. Women Professors, 1997/1998.

Level	Total	% Women
All Fields		
Full professors and professors	47,944	30.9
Professors under temporary contract	31,103	32.7
Assistant professors	6,874	42.5
Total	85,921	33.3
Physics Only		
Full professors (Catedráticos)	443	2.9
Professors (Titulares)	1,488	25.2
Professors under temporary contract (Asociados)	422	21.7
Assistant professors	255	27.9
Total	2,608	21.0

In 1939 the *Consejo Superior de Investigaciones Científicas* (CSIC, www.csic.es) was founded using the facilities of the former JAE, with the purpose of undertaking scientific and technological research in all branches of knowledge. Today the CSIC has a permanent staff of around 4500 persons arranged as follows: 2000 persons are scientific staff, 1800 persons are support staff, and 700 persons work as administrative or in other positions. In addition, 2000 persons are scientific training staff. The general ratio of women in the scientific staff is 30%. The ratio of women on the executive board is around 19%. The scientific staff is divided into three levels. The overall ratio of women at the highest level is only 13.3%, but this proportion decreases in the areas related to physics, as can be seen in Table 2.

The ratio of female members is 19% in the Spanish Physical Society (*Real Sociedad Española de Física*, RSEF; www.ucm.es/info/rsef). Although the society is open to all physicists, the members are mainly university professors and researchers. From a total of 53 members on the board of the society, six (11%) are female and only two of them have a responsibility charge: the general secretary and the editor of the society journal. To attract students to study physics, the RSEF organizes special programs (e.g., Physics on Stage) coordinated with

CP628, *Women in Physics: The IUPAP International Conference on Women in Physics,* edited by B. K. Hartline and D. Li
© 2002 American Institute of Physics 0-7354-0074-1/02/$19.00

TABLE 2. Women Staff Members in the Physical Sciences and Technology at the *Consejo Superior de Investigaciones Científicas*.

Staff Level	Men	Women	Total	% Women
Professor of research	92	5	97	5.2
Research scientist	105	29	134	21.6
Scientist	228	114	342	33.3
Total	427	149	576	25.9

several European institutions, including the European Laboratory for Particle Physics (CERN), the European Space Agency (ESA), and the European Southern Observatory (ESO).

Other professional societies have a more significant female participation. The Graduate Association (*Colegio de Licenciados*), for professionals involved in teaching at the secondary-school level, has a higher female ratio than other professional organizations, because this is one of the professions that attracts a higher number of women. We estimate that about 50% of physics teachers at the secondary level are women.

Medical physics is another area of significant participation by women in Spain. Of 345 members of the Spanish Society of Medical Physics (www.sefm.es), 29% are women. The ratio of women at the highest level (heads of Medical Physics Departments) is 28%.

There are enormous difficulties in finding good gender-desegregated statistics, data, or general information concerning the female rates in industry in our country. Women at the top levels of companies (managers/presidents or department heads) are fewer than 10%. In the report from the ETAN Expert Working Group on Women and Science it is estimated that the percentage of female managers/presidents of Spanish companies taking part in research projects is around 4%.[1] The lack of gender-divided statistics is a general problem in Spain because most of our institutions do not deal with the status of women at work. In 1983 the Women's Institute was created (Ministry of Work and Social Affairs, www.mtas.es) to defend women's rights and promote gender equality. Recently, the first divided-gender statistics ("*Mujeres en Cifras*") were published by the institute.

REFERENCE

1. European Commission, "Science Policies in the European Union—Promoting Excellence Through Mainstreaming Gender Equality," a report from the ETAN (European Technology Assessment Network) Expert Working Group on Women and Science (Office for Official Publications of the European Communities, Luxembourg, 2000); available in English, French, German, and Italian at http://www.cordis.lu/improving/women/documents.htm.

Women in Physics in Sudan

Mai Osman

Very few women in Sudan were allowed to go to school in the early days. This was the case up to the early 1950s. After that time, more and more women had the chance to get an education as more girls' schools were opened. Now we can correctly assume that 60% of university graduates are women.

The reason behind this is that Sudanese families are more dependent on men to secure living expenses and to make ends meet, which leaves women with a better chance of climbing up the educational ladder.

When discussing physics in Sudan, we can safely say that physics is the last choice for applicants at all universities. The reason is simply that there are very few careers in physics because there are very few research centers in Sudan. And for those, physics is not a priority. So the best career choice for almost all physics graduates is teaching at schools and universities. Because teaching is not a well-paying job, many women graduates tend toward postgraduate studies so they can find employment at a university with better pay.

Sudan is considered to be one of the first countries in Africa and the Arab world where education began relatively early. It has become a part of our modern culture for women to get as much education as they can in physics as well as in other subjects. But for physicists, the problem is one of finding employment where they can experience job satisfaction and make full use of their qualifications and knowledge.

Funds are needed from international organizations and well-off countries for research centers in Sudan. This will help a greatly in improving the careers of women physicists and give them more satisfaction and joy. Utilizing the knowledge, qualifications, and experience of women physicists in Sudan will contribute much to the world of physics.

CP628, *Women in Physics: The IUPAP International Conference on Women in Physics,* edited by B. K. Hartline and D. Li

Swedish Women in Physics

Lotten Glans[1], Cecilia Jarlskog[2], Björn Jonson[3],
and Katarina Wilhelmsen Rolander[4]

[1]*Mid Sweden University;* [2]*European Laboratory for Particle Physics (CERN) and Lund University;*
[3]*Chalmers University of Technology and Göteborg University;* [4]*Stockholm University*

A few Swedish studies were done in the 1990s by researchers in Umeå on women physicists and chemists in Sweden and the career paths of women in engineering science. The data on Swedish female physicists span 1900–1989 and include when they received their Ph.D. and in which areas. Only a handful of women defended their theses at the beginning of the century, and another handful between 1930 and 1970. From the 1970s on, the number increased. In total, 59 women in Sweden obtained a Ph.D. in physics from 1900 to 1989, which is 8% of the total number of physics Ph.D.'s awarded. During the same time period, 213 women (or 21%) obtained Ph.D.'s in chemistry.

INCREASED PARTICIPATION

The number of both undergraduate and graduate students in physics in Sweden is currently about 30%. At higher levels, though, the number is significantly smaller. Recent statistics indicate that 7% of Swedish professors overall are women. The percentage of women in natural sciences is smaller than in other subjects.

As an example, at Stockholm University the official equal opportunities policy clearly states that the balance between the genders of the university teachers should be between 40% and 60%. Today, 11% of the tenured staff at Stockholm University are women. One could argue that this reflects many decades of past sins; however, if one takes a closer look, the number of women among the graduating Ph.D.'s is not reflected when it comes to hiring new staff. Thus, the actual situation is far from the goal set by the authorities, and there is a lack of policy on how to increase the number of women holding senior positions.

To increase the number of women in physics is important for the same reason it is important to increase the number of women and other minorities in every field. To have a heterogeneous working group means variety in how to approach and solve problems, which questions are asked, and which ideas come up.

FEATURES OF CAREER

The features of an academic career in Sweden may be peculiar, or at least specific to, the Nordic countries. However, the career for a woman in physics does not differ from a career in engineering, chemistry, etc. The typical career path within the universities is as follows. After an undergraduate education (around four years), one applies for a Ph.D. position. If the student does research full time, the Ph.D. should be achieved within four years. Normally such a position is 80% research and 20% teaching and the Ph.D. is completed in five years. After this it is very common (and in most cases recommended for a future career) to apply for a postdoctoral position abroad for one to two years. Next, most people apply for a time-limited position as a junior researcher (forskarassistent) for four to five years, where most perform research 80% of the time. The next step is as senior lecturer, a permanent position with more teaching than research (in most cases), and thereafter a full professorship. Of course, there are deviations from this path, with some full-time research positions and full-time teaching positions.

In Sweden every family has a right to 12 months of parental leave at 80% of one's salary. This time can be shared between the parents, but one month is reserved for each parent and cannot be transferred to the other. There is also a legal right to child care for every child between 1 and 12 years of age. In reality the career path of women seems to be more affected by having a family. In practice, most parental leave is still maternity leave.

CP628, *Women in Physics: The IUPAP International Conference on Women in Physics,* edited by B. K. Hartline and D. Li

SERIOUS BARRIERS

The most serious barrier for women in physics is the career path, or rather the lack of one. After each time-limited position one has to apply for a new type, with absolutely no guarantee of obtaining one even if one has done well. Instead, factors such as the economy of the universities, the number of teachers close to retirement, and contacts and support from older colleagues are important. When younger women start to be real competitors they are not supported in the same way as young men are. The Swedish career path at universities also means that a permanent position is seldom obtained before the age of 35, in sharp contrast to industry. For women, and an increasing number of men today, this is a reason for not choosing an academic career, or even for not getting a Ph.D. Economic instability is not compatible with a family and children. Also, the weight put on a postdoctoral position abroad prohibits women from choosing a continued academic career.

JOYS/SATISFACTIONS

Most of the joys and satisfactions are not due to the fact that one is in physics, but rather that one works in research and teaching. For instance, it is rewarding to meet and interact with students and colleagues with different backgrounds. Researchers in most areas get the chance to meet different people and exchange ideas. Most teachers/researchers also work in an area they have chosen because it is what they are interested in and like best.

ATTRACTING WOMEN

Tenure earlier seems to be an obvious way to go and to increase the number of women at the physics departments. A clear policy when hiring new staff would be beneficial. Before announcing a position, care should be taken to ensure that there exist highly competent women that might be likely to apply for the position. If there are no female applicants the board should reassess the field of interest for the position in such a way that attracts female physicists of high competence to the department.

In 1996 the Government proposed 30 professorships for competent women as an affirmative action. A number of junior research (forskarassistent) and Ph.D. positions for women were also proposed. There was, however, some ill feeling concerning the professorships and a tendency to hint that "Had these women been good enough they would have gotten positions in competition with men." This was a one-time measure and Swedish law prohibits positions for one gender only. If the qualifications are equal, gender may be used as a factor in the selection.

Locally at universities and schools initiatives are taken targeting girls, e.g., female physics students visit schools and high schools to attract girls to science.

In 2001, a conference, "Women Talking Physics" was held in Sweden at Uppsala University. A network of female physicists was formed and a second conference is planned for May 2002. One of the contributions to that meeting will be a report from this conference.

Forming a new section called "Women in Swedish Physics" has been suggested to the Swedish Physical Society. This would be a forum for women physicists from different universities and disciplines to meet and share ideas about physics and also share experiences and knowledge on how to get ahead in a male-dominated profession.

Women in Physics in Switzerland

Iris Zschokke-Gränacher

University of Basel

Recently a physics institute at a Swiss university had the idea of organizing an "Open House for High School Girls Only" day. It turned out to be a great success—many more girls came to see the laboratories than were expected. What better could one imagine than high school girls going home and telling their teachers and parents that they want to learn more about physics!

In Switzerland the percentage of female students in physics has increased from 9% to 13% over the last 10 years, while the total of physics students has declined by 10%. Of those who earned their doctoral degree in physics in 2000, 8.6% were women (Germany: 9.8%). This corresponds to an increase of 3.8% over the last decade. Of the full professors in experimental and theoretical physics, 5% are women.

The Swiss National Science Foundation, the universities, and the Swiss government offer various programs and fellowships to encourage women in research and in academic careers. As a result, the number of female postdocs in physics has increased considerably in recent years.

As an incentive for women who want to resume research after some years' interruption for family reasons, the Swiss National Science Foundation has founded a special grant. At least 10 woman physicists ages 30 to 45 have benefited from the Marie Heim-Vögtlin Fellowship, named after the first female medical doctor in Switzerland.

Besides these diverse initiatives on the university level, it is of great importance for the research system itself that women are represented on national and international committees for the planning, evaluation, and management of science and education. In order to find and contact women with extensive experience in a particular field, a Swiss database of woman scientists and experts has been established: www.femdat.ch.

Twenty years ago the first and, until now, the only woman ever elected for president of a Swiss university was Verena Meyer (University of Zürich), a professor of experimental physics. This was a real breakthrough that attracted the attention of the scientific community as well as the public in Switzerland. It reinforced that physics can be suitable for women!

Today in our country, almost every woman professor in physics holds a leading position on an international or national expert committee, such as:

- Board Member of the Quantum Optics Division of the European Physical Society (Ursula Keller, Swiss Federal Institute of Technology Zürich, European Laboratory for Particle Physics [ETHZ]).
- Chairperson of Executive Board for Compact Muon Solenoid for the Large Hadron Collider at CERN (Felicitas Pauss, ETHZ)
- Project Manager for the ROSINA/ROSETTA Experiment of the European Space Agency (Kathrin Altwegg, University of Bern)
- Member of Municipal Parliament of the City of Geneva (Christina Matthey, CERN)
- Member of the board of the Swiss Federal Institutes of Technology (Iris Zschokke)

More could be done to encourage young female physicists to become more self-confident in applying for their own research funds. To accomplish this, the Swiss National Science Foundation is reviewing various possibilities such as providing extra information and assistance for women who are first-time applicants for a research grant. Another problem is the availability and quality of child-care centers at universities and research institutions.

The most serious problem in our country, however, is the quality of physics education at the high-school level. There is a real need for improvement, especially on the political level. However, initiatives such as those discussed above might be of great help, and this would not only improve education but also point out to future generations the beauty and omnipresence of physics in their lives.

CP628, *Women in Physics: The IUPAP International Conference on Women in Physics,* edited by B. K. Hartline and D. Li
© 2002 American Institute of Physics 0-7354-0074-1/02/$19.00

Women in Physics in Tunisia

Zohra Ben Lakhdar[1], Souad Chekuir Lahmar[1], and Menana Kilani Gabsi[2]

[1]Faculté des Sciences de Tunis; [2]Lycée El Omrane

Tunisia is a young state with an age-old civilization. Founded over a thousand years before the Christian era by Elissa, a woman from Tyr, Carthage once shone over the entire Mediterranean Basin. Because of its geographic location, Tunisia experienced countless foreign conquests (Phoenicians, Romans, Berbers, Arabs, Spaniards, Turks, and the French Protectorate from 1881 to 1956). Tunisia is a pure Mediterranean country with a multiplicity of civilizations.

Tunisia obtained its independence from the French Protectorate on March 20, 1956. Its population was 9.3 million in 1998, having grown from 3.5 million in 1956.

Once independence was acquired, Tunisia's social policy was aimed at modifying the country's overall social structure—reforming agriculture, health, housing, education—with one historical first measure: the family statute. On August 2, 1956, the family statute was passed, granting women their rightful place as equal citizens. As a result, Tunisian girls are able to attend school freely—girls and boys are in the same school—with interaction at all levels. Primary and secondary school teachers are now almost 50% women; at the university 30% are women. Table 1 shows the dramatic contrast in the ratios of women students before and after the family statute was enacted.

TABLE 1. Evolution in the Number of Girls and Boys in School.

Year	Primary School		Secondary School		University	
	1954-55	1998-99	1954-55	1998-99	1954-55	1998-99
Girls	49,000	679,000	1500	416,500	–	–
Boys	177,736	754,000	16,000	416,500	–	–
Rate (% girls)	22%	47%	8.5%	50%	–	48%

WOMEN IN SCIENCES

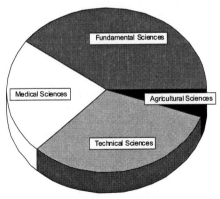

Total Science Students = 137,024

Figure 1. Distribution of science students, 1997-1998.

The total number and distribution of students in sciences in 1997-1998 is shown in Figure 1. The gender distribution of students in sciences in this same timeframe was 54% males and 46% females, with 60%–70% of the women enrolled in medicine and biology and 30% enrolled in physics.

In Tunisia, because of the limited number of spaces, the choice of university studies (engineering, medicine, physics, or human science) is not free, but depends on the results obtained on final exams (baccalaureate) in secondary school. The first choice is for medicine, chemistry (pharmacie), or IHEC (Institut des hautes études commerciales); the last choice in sciences is for physics at the Faculty of Sciences. The best results are obtained by girls; as a result, the number of female students in medicine, chemistry, and biology is actually greater than male students. In 1997, it reached 75% in the first year of medicine, then decreased with years of study.

CP628, Women in Physics: The IUPAP International Conference on Women in Physics, edited by B. K. Hartline and D. Li
© 2002 American Institute of Physics 0-7354-0074-1/02/$19.00

Table 2 lists the percentage of women/men who received a diploma in science in 1996-1997 at the largest university in Tunisia.

TABLE 2. Science Diplomas Awarded at the University of Tunis, 1996-1997.

	Technicians	Medicine	Maitrise (Bachelor's Degree)	Engineer
No. of Women	363	84	440	139
No. of Men	587	153	1225	670
% Women	38%	35%	27%	17%

In the Job Market

Males and females with the same qualifications are treated equally in the job market. There is no discrimination in getting a job. There is also no difference in salary for the same level.

In Practice

At home women are not yet equal to the men. Women are considered the primary caregivers in the family. As a result, women look for jobs that offer the same holidays as children, where they have less responsibility and less preparation at home. Family responsibilities limit the ambitions of women for developing careers. This situation is also encouraged by society. Women are more numerous in teaching positions and in public jobs. In a family we expect that the man earns more and has the better job.

Table 3 shows the percentages of women in two different age groups who worked between 1975 and 1997.

TABLE 3. Percentage of Women in the Workforce, 1975–1997.

Year	25–29 years old	30–34 years old
1975	21.2%	16.2%
1984	29.3%	23.4%
1997	36.3%	29.0%

WOMEN IN PHYSICS

The number of female physics students is 30%, assistants 35%, professors 30%, and directors of research laboratories 5%. Women at universities and research centers have the same amount of office and research space as men.

The number of students in physics, both male and female, is decreasing . Between 1997 and 2001, the number of physics students decreased by 50%—the same percentage for girls and boys. Students are studying information technology, computer sciences, and biology because, they say:

- No jobs exist for graduates with a master's in physics.
- Physics courses are difficult and boring.
- Physics is not introduced in the culture, unlike information technology, computer science, and biology, which are present in the environment and reinforced by mass media and by products on the market.

Physics is introduced theoretically, traditionally, without creativity, and without culture at a very late age (14–15 years). Women, scientific societies, and others can work together to make physics more attractive.

REFERENCES

1. "Tunisia works," Secretariat d'Etat à l'Iformation, Gouvernement Tunisien (1960).
2. Jeune Afrique, n°1837, 20-26 (Mars 1996).
3. A. Beschaouch, *La Légende de Carthage*, Collection Découvertes Gallimard No. 172 (Gallimard, 1993).
4. Le choix de l'avenir, Ministère des télécommunications (Octobre 1999).

Statistical Distributions of Women Physicists in Turkey

Saziye Ugur[1], Engin Arik[2], Ayla Celikel[2], and Demet Kaya[1]

[1]Istanbul Technical University; [2]Bosphorous University

Turkey is a Middle Eastern country that lies mainly in southwestern Asia, with a small area extending into southeastern Europe. It is a bridge between Asia and Europe.

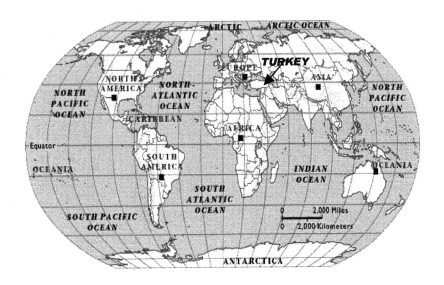

Population of 70 million:
Ages 0–24 55%
Ages 25–64 41%
Ages 64+ 4%

Surface area 815 km^2
Men/ women 50%-50%
Reading-writing rate 90%
City/village 41%-59%

Turkey was proclaimed after World War I by Mustafa Kemal Atatürk on October 29, 1923. Atatürk introduced reforms that he considered of vital importance for the salvation and survival of his people between 1924 and 1938. These reforms were welcomed enthusiastically by the Turkish people, especially by the Turkish women. In 1934 women were given the right to vote for deputies to the National Assembly, and to run for office. In this matter, Turkey was ahead of many European nations.

Among the Ottomans (14th century through early 20th century) there were many madrashahs, where Islamic thought was the subject of study; only in 1900 was the first university opened in İstanbul. True, universities were needed if science and letters were to progress, and in 1933 Atatürk founded the University of İstanbul. This is an important date for higher education in Turkey.

Turkish women work in all fields today, especially in science. There are proportionately more women physicists in Turkey than in Europe and America. The distributions of Turkish women in physics among and within universities are shown in Tables 1 and 2, respectively.

CP628, *Women in Physics: The IUPAP International Conference on Women in Physics,* edited by B. K. Hartline and D. Li
© 2002 American Institute of Physics 0-7354-0074-1/02/$19.00

TABLE 1. Women Physicists on University Faculties in Turkey, 2001.[a]

Universities	Percent
Istanbul Technical University	41
Bosphorus University	50
Middle East University	20
Yıldız Technical University	44
Marmara University	31
Istanbul University	50
Uludağ University	21
Trakya University	28
Ege University	42
Dokuz Eylül University	40
Ankara University	30
Hacettepe University	42
Gazi University	27
Balıkesir University	28
Dicle University	35
Fırat University	7
İnönü University	14
Anadolu University	48
Atatürk University	17
Harran University	38
Mersin University	25
Muğla University	37
Sakarya University	17
Selçuk University	25
Süleyman Demirel University	19
Yüzüncüyıl University	15
Zonguldak Karaelmas University	46
Cumhuriyet University	38
Osmangazi University	31
Erciyes University	25
Çukurova University	33
Dumlupınar University	10
İzzet Baysal University	8
Celal Bayar University	25
Osman Gazi University	31
Fatih University	9
Teditepe University	27
Average	29

[a] Professors, assistant professors, associate professors, and other ranks.

TABLE 2. Faculty Positions in Physics Departments held by Women, 2001.

Academic Rank	Percent
Full Professor	11
Associate Professor	12
Assistant Professor	12
Ph.D.	3
Lecturer	4
Researcher	5
Research Assistant	24
Ph.D. and Master's	19

- In 2001, 41% of university students were female.

- In the 1999-2000 academic year, 40% of the total graduate students were female in physics departments at all universities. According to the Turkish Student Selection and Placement Center, this percentage has decreased to 34%.

- The Council of Turkish Higher Education reports that in 2001, 25% of professors overall were female.

According to the results of the 2001 survey by the IUPAP Working Group on Women in Physics:

- 51% of women physicists are single in Turkey. 47% of married women in physics have children.

- The majority of married women were affected by their marriage in a good manner, while just 11% said they were negatively affected by their marriage. Yet having a baby plays an important role in a woman physicist's scientific life, and large amount of physicist-mothers said that they preferred giving up their jobs when they had to make a choice.

Women in Physics: Ukraine and Global Trends

O.V. Patsahan

Institute for Condensed Matter Physics of the National Academy of Sciences of Ukraine

Is physics an exceptional privilege of men? The answer to this question has already been given by women themselves. The role of women in the society at large and in science in particular has grown considerably in the last decades. The past century was characterized by great discoveries in the field of physics and by its rapid development. And women have made a valuable contribution to this progress. The names of M. Sklodowska-Curie, M. Goeppert-Mayer, and L. Meitner are well-known throughout the world.

While at the beginning of the 20th Century women physicists were rare bright personalities whose overwhelming desire to comprehend the secrets of the surrounding world overcame the established public views about a home predestination for women, in the second half of the century a woman in a physics laboratory or on the physics faculty of a university was a common sight.

General global trends show that for the several past decades the number of women studying physics at higher educational institutions and working for a Ph.D. in physics has increased in all countries. However, the quota of women occupied in physics is still significantly lower than that in chemistry or biology. Also, the number of women physicists decreases essentially with each step up the academic ladder.

The situation in Ukraine is similar (see Figures 1 and 2). But, along with the factors typical of most countries—raising women's level of education, gaining financial independence from men, emancipation from housekeeping—Ukraine has some additional factors not observed in highly developed countries. On the one hand, Ukraine, like other European states, has a deeply rooted historical tradition of equal rights and opportunities for men and women in choosing their professions. In Ukraine women have always been an active part of the society, and the number of men and women with higher education is approximately the same here.

On the other hand, Ukraine is undergoing a complex process of building a democratic state, which causes great changes in its economy and affects its economic situation. At present, Ukraine is one of the poorest countries in Europe. This factor has a strong impact on the development of Ukrainian science on the whole and of physics, in particular.

In fact, at the beginning of the 1990s, Ukraine possessed a powerful scientific potential in physics. For the last decade the situation has changed for the worse. The insufficient financing of science forced many Ukrainian physicists to emigrate abroad or go into business. However, these generally negative factors caused an influx of

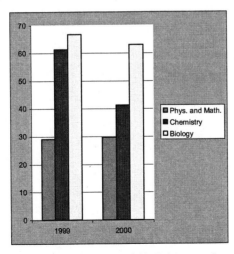

Figure 1. Percent of Ph.D.'s earned by women in selected fields, 1999–2000.[1]

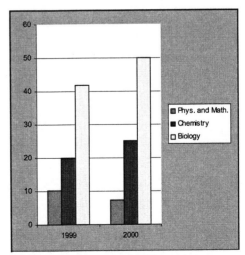

Figure 2. Percent of Doctor of Sciences degrees earned by women in selected fields, 1999–2000.[2]

CP628, *Women in Physics: The IUPAP International Conference on Women in Physics,* edited by B. K. Hartline and D. Li

women into physics in Ukraine. Thus, the number of women among Ph.D. students, researchers, and university students in physics has grown. And this is not only due to the lessened competition (which is also an essential factor). This tendency shows that women can do difficult scholarly work that often does not yield quick results and has been considered men's prerogative.

In spite of these favorable-at first-sight contemporary conditions for women physicists in Ukraine, women still face obstacles. One of the major problems hindering women's professional and career advancement is balancing the family and career.

A nuclear family is still common for the Ukrainian society. A woman often plays the principal role in the family: she works and provides her share of the family income, takes an active part in raising the children, and does the major part of the housekeeping. The latter takes a lot of time and largely depends on the financial situation of the family. Therefore, professional and career advancement requires more mental and physical effort from women scientists than from men. A woman often faces the choice: family or career. As a result, there is a high percentage of single and childless women among those occupied in physics.

It is common knowledge that the lower the level of a society's economic development, the less time its women have for raising their educational and cultural levels. The present economic situation in Ukraine does not give grounds for an optimistic prognosis in this respect. However, one can assert today that despite difficulties, women dispel the myth about men's reign in physics.

REFERENCES

1. Scientific World, Nos. 1-12 (1999), Nos. 1-12 (2000) (in Ukrainian).
2. Bulletin of Highest Qualification Committee of Ukraine, Nos. 1-6 (1999), Nos. 1-6 (2000), No. 1 (2001) (in Ukrainian).

The Current Status of Women in Physics in the United Kingdom

Gillian Gehring[1] and Ann Marks[2]

[1]University of Sheffield; [2]Physics Teacher

In 1994 the Minister for Science, William Waldergrave, said, "Women are the country's single most underused resource." In the UK the proportion of undergraduate women in physics is low compared with many countries, and the number of postgraduate women is swelled by the influx of students from abroad. Significantly fewer female university students study physics and engineering than study mathematics or chemistry.

Physicists as a whole have very worthwhile careers and we believe that there are many more women who could benefit from following a physics education. The opportunities are there. A number of research laboratories with mixed workforces show that women make a special contribution to the working environment. It appears that organizations employing 20% or more women see benefits not only for the women but also for the men.

CONTRIBUTIONS OF WOMEN PHYSICISTS

The UK is a country with a large number of very distinguished women physicists. A few are mentioned below. Astronomy and crystallography have attracted a relatively large fraction of the UK women physicists.

- *Jane Brown* is a neutron crystallographer who has spent most of her working life at the Institut Laue Langevin in Grenoble. She has been awarded the 2001 Walter Halg prize by the European Neutron Scattering Association; this prize is awarded biannually.
- *Jocelyn Bell-Burnell* identified the first object in the sky that was finally identified as a pulsar when she was a graduate student. This earned her supervisor a Nobel Prize.
- *Athene Donald FRS* is one of the country's leading polymer physicists. She is the first woman professor in the Cavendish Laboratory at Cambridge.
- *Elizabeth Gardner* was a Scot who made seminal discoveries in the theory of neural networks. This was done in the extremely short time available to her, as she died at age 29.
- *Carole Jordan FRS* is the UK's leading expert on the ultraviolet and x-ray spectra of cool stars, including the sun. She made the first identifications of many important emission lines, and has used a combination of observations and theory to understand stellar chromospheres and coronae.

BARRIERS

The working environment in the UK is such that women have many barriers to overcome if they are to pursue successful physics careers. These barriers may be listed briefly as follows:

1. The British have a culture of long working hours, which could be regarded as counterproductive.
2. Academics usually gain their first permanent position in their late 30s following up to 10 years spent on short-term contracts, during which it is common to change institutions and even countries. Most women are not mobile by their mid-30s. Because women are usually younger than their partners (who are frequently physicists), they are the more junior in their careers. More often their careers suffer, particularly when a move is necessary. Women in academia either work full time through their childbearing years or delay starting a family until they have secured a permanent post. Often the talents of those who are not happy with either of these options and take a career break are lost or underused.
3. The UK has a mobile workforce; family and friendship ties are fragmented, making child care difficult.
4. It is rare that working part-time and/or job sharing are successful in developing a career. It is not part of the culture in a research or university environment.
5. There has been a steady destruction of industrial research centers over the last 30 years and the number of government research centers has also been considerably reduced. A number of physics departments

CP628, *Women in Physics: The IUPAP International Conference on Women in Physics,* edited by B. K. Hartline and D. Li

within universities have also closed. This has led to a situation in which we believe that there are fewer tenured physicists concentrated in fewer centers in the UK.

6. There is a culture of mini-insults and sexist jokes. Over a long period these undermine confidence.

IMPROVEMENTS AND INITIATIVES

One could have a mighty bonfire—and crash a computer!—with all material on why more women do not practice science. Below we list the initiatives that in our opinion may have done some good.

Initiatives that cover all science engineering and technology specifically to assist women:

- The government commissioned the Rising Tide Report in 1994 (ISBN 1994 0 11 4300968). A number of initiatives have followed, most notably a unit in the Department of Trade and Industry (www.set4women.gov.uk) and an umbrella organization to disseminate all science and engineering information (http://www.awise.org/) .
- Athena (http://www.athena.ic.ac.uk/) was founded by the Higher Education Funding Council to advance women in science, engineering, and technology in higher education. It provides funds to encourage universities to increase the recruitment, retention, participation, progression, and promotion of women.
- The Daphne Jackson Trust (www.daphnejackson.org) assists women returning to science careers after a break by offering two-year part-time fellowships for retraining. The Wellcome Trust (www.wellcome. ac.uk) has an excellent program for those whose work is related to medicine or dentistry.
- The Royal Society introduced the Dorothy Hodgekin fellowships, designed to combat the problem of mobility by introducing four-year fellowships that the holder can take to any institution. Only 15 are awarded per year for all science and engineering, and almost all go to women. The Wellcome Trust also has fellowships in this area, again restricted to those in medical fields.
- Women have started web sites for mutual support and to share information, e.g., Portiaweb (www.portiaweb.org) and Daphnet (daphnet@ic.ac.uk).

Initiatives that have increased possibilities for women as a by-product:

- The research councils, EPSRC and PPARC, and the Royal Society have introduced five-year fellowships. These are useable as "starters" to permanent posts and have often enabled first-rate women scientists to obtain a post at a university of their own choosing.
- The National Curriculum introduced physics into the core curriculum for ages 5 to 16. This has dramatically increased the number of girls studying some physics up to age 16.
- The new, more flexible first-year sixth-form courses for those ages 16/17 should increase the number of girls taking at least some specialized physics.
- In the UK the financial service and IT industries are booming. These two areas take all the numerate graduates they can find. This has meant that trained women scientists have had increasing possibilities of returning to work in a nonscientific career after a break.

Physics schemes:

- Universities physics departments have tried to boost the number of women studying physics by holding special open days for schoolgirls.
- The Institute of Physics *Career Breaks* leaflet and career development workshops have been of benefit.

THE FUTURE

We are proud that some women are getting through the glass ceiling. The number of women professors of physics has increased from 1 to over 10 in about 10 years and a few women with physics qualifications are appearing in other top jobs. A sizeable number of women in top positions have managed to combine a successful career with a full family life. However, we need to tackle the appalling waste of female talent that certainly occurs because of the barriers mentioned earlier. We also want to learn how to increase the number of women embarking on a study of physics.

We intend to press for allowable conference expenses to include the extra cost of child and elderly care required when a caregiver of either gender is away. In both industrial careers and academia, people are at a disadvantage when they cannot attend conferences.

ACKNOWLEDGMENTS

We thank all members of the UK Women in Physics Group and particularly Liz Whitelegg for her input.

Women in Physics in the United States

Megan Urry[1], Sheila Tobias, Kim Budil[2], Howard Georgi[3], Kristine Lang[4], Dongqi Li[5], Laurie McNeil[6], Peter Saeta[7], Jennifer Sokoloski[8], Sharon Stephenson[9], Aparna Venkatesan[10], and Yevgeniya Zastavker[11]

[1]*Yale University;* [2]*Lawrence Livermore National Laboratory;* [3]*Harvard University;*
[4]*National Institute of Standards and Technology;* [5]*Argonne National Laboratory;*
[6]*University of North Carolina;* [7]*Harvey Mudd College;*
[8]*Harvard/Smithsonian Center for Astrophysics;*
[9]*Gettysburg College;* [10]*University of Colorado;*
[11]*Wellesley College*

In the last 25 years, women in the United States have made great progress in expanding their participation in professions formerly populated almost exclusively by men. However, this progress has been much more limited in physics than in many other fields. The physics community in the United States has made slow progress in enrolling and rewarding women in physics compared with other professional communities, due in part to the belief that because science is an "objective" pursuit, the underrepresentation of women is simply an indication of their lack of interest or ability in the field, rather than an indication of discrimination or exclusion.

Thus, unlike other countries that have looked to women as a resource to increase their number of physicists, the United States could for most of the 1940s, 1950s, and 1960s accept as "natural" the tendency of women students—even those with outstanding mathematical talents—not to select physics as a subject of study. It was not until the mid-1960s, when the women's movement exposed the disproportion of women in physics and took action to eradicate discrimination and to eliminate barriers, that anything was done to address what was until then not considered to be anyone's problem—neither women's nor the country's.

There were always exceptional women physicists in the United States, as elsewhere. But here in the United States they were not just exceptional, they were rare and solitary. Even when they were "successful" by ordinary measures, they did not thrive at the same rate and to the same extent as men. Many did not marry or have children. Most remained research associates, off the professorial ladder and, if they taught at all, taught at women's colleges. As a result, their presence in the physics world—scant and segregated as it was—did not counter the overall impression, expressed by physicist I.I. Rabi as late as 1982, that women do not have the temperament to do physics. "They may do science," Rabi told an interviewer toward the end of his life, "but they will never do Great Science."

Twenty years later, according to "Women in Physics, 2000," a report from the American Institute of Physics, women remain "sorely underrepresented" in physics in the United States.[1] They earn fewer than 20% of the bachelor's degrees and fewer than 12% of the Ph.D.'s, despite significant strides for women in the other sciences (50% of the bachelor's and 30% of the Ph.D.'s in chemistry, 50% of both in biology).

Isolation for young women who do select physics is one result of this disproportion; a sense of being "out of place," another. Only 20 departments of physics in the United States graduate five or more women physics majors per year, which means that in most universities women students are still an unusual occurrence, reinforcing stereotypical views among senior physicists and the women's peers that women still can't or won't do physics.

Efforts in the United States to correct this (led by the physics community itself) have focused on three issues: the "culture" of physics in the United States, which encourages a hyper-competitive, "overly masculinized," almost monastic approach to science; work-life issues that penalize young people (not just women) who have working partners and children; and perceived abuse, which ranges from being discounted and not taken seriously by professors, employers, and colleagues to being openly discouraged, disliked, and having one's career intentionally derailed.

Since the 1970s, the American Physical Society (APS) has provided information and support in the struggle to increase the numbers and success of women in physics. In 1972, the society established a Committee on the Status of Women in Physics, both to study the problem and to try to do something about it. Led by senior women in the field, the committee collected and published detailed information about women in physics and made certain that women were mentioned and earned their share of APS honors, given visibility in organizing research sessions, and given high-level appointments in the society itself. The committee has also prepared materials such as posters and leaflets to help in recruiting women to physics by highlighting women's contributions to the field, and has supported in-career women by sponsoring scientific sessions and providing networking opportunities at major society meetings.

Perhaps their most far-reaching effort is the committee's well-publicized Climate for Women Site Visit Program to academic physics departments, in which teams of senior female physicists interview women students (graduate and undergraduate), professors, and other staff. Armed with their findings, the Site Visit Team then conducts an "exit interview" with the department chairman and prepares a report outlining a plan for improvement. This effort grew out of a request in the late 1980s from the chairs of physics departments for advice on how they could recruit and retain more female students and faculty. The National Science Foundation funded the initial visits and the program continues today with funding from APS. As of March 2002, 27 such visits have been conducted, including two to national laboratories and one return visit to an academic department.

Other physicists aligned with women activists have tackled:

1. The shortage of women teachers of physics at all levels of schooling in the United States
2. Problems of perception and bias in the evaluation of women's work
3. Sexual harassment
4. Physical safety for women in labs
5. The difficulty of managing marriage to a professional partner and their own professional career

A sixth issue is childcare, which in the United States is typically not supported either by the state or by the employer. Ours is a country where in-home help is uncommon in the middle classes, which leaves the burden of maintaining a household on the family itself.

Americans, like physicists elsewhere, want to see more women in physics. There is the fairness issue, but along with that there is a pressing need for talent in physics, wherever it may be found. The socialization of women and their adult roles cause them to bring fresh perspectives to the selection of research problems, to the organizing of research groups, to the teaching of physics, and to the work of physics itself.

Because the United States prepares and trains the future research faculty of many other nations, our country's ability to recruit and retain women in physics directly affects the status of women in physics elsewhere, making the questions we bring to IUPAP even more urgent: What keeps girls interested in physics? How do we counter residual bias against women in the field? What works in attracting and supporting women students and women professionals? What constitutes a critical mass, meaning a stable cadre of women physicists that will persist and prosper? And how do we get there?

REFERENCE

1. R. Ivie and K. Stowe, "Women in Physics, 2000," AIP Publ. No. R-430 (American Institute of Physics, College Park, MD, June 2000); www.aip.org/statistics/trends/highlite/women/women.htm.

Women in Physics in Uzbekistan

Ulmas Gafurov

Institute of Nuclear Physics, Uzbek Academy of Sciences

HISTORY

In our land (Middle Asia) reside the richest and oldest fundamentals of cultural and scientific heritage. The greatest contributions to world culture and science have been made by widely known outstanding scientists such as al-Khorezmiy (8th and 9th centuries; the mathematical term "algorithm" was named after him), al-Farabiy (9th and 10th centuries), Abu ibn-Sina (known in the West as Avicenna; 10th and 11th centuries), Abu Rayhan al-Biruni (10th and 11th centuries), Omar Khayyam (10th and 11th centuries), and Mirzo Ulugbek (14th and 15th centuries). It was Biruniy who first suggested the Earth rotated around the sun, long before Western scientists came to that conclusion. Omar Khayyam developed third-range nonlinear algebraic equations. Ulugbek built the famous observatory and determined the coordinates of some planets and many stars.

Uzbekistan enjoys a nearly 100% literacy rate and its people generally have a middle-school education.

MODERN DEVELOPMENT OF PHYSICAL SCIENCE

The modern development of physics and other natural sciences is connected with the arrival in Uzbekistan of large groups of scientists from central Russian cities in 1918, and the founding in Tashkent of the Middle Asian State University, now National University, following the creation of the Uzbek Academy of Sciences in 1943.

At present there are six physical science research institutes in the Tashkent, Karakalpakistan, and Samarkand branches of the Academy of Sciences of Uzbekistan. Among these institutes is the largest in Central Asia, the Institute of Nuclear Physics, founded by Sergey Starodubcev and Ubay Arifov in 1956. Professors Starodubcev and Arifov also founded the Physical-Technical and Electronics institutes. Many of their apprentices and followers (including many women physicists) are working in Uzbekistan and other countries.

WOMEN RESEARCHERS

Research at the Uzbek Academy of Sciences institutes and universities covers a wide range of physical fields: nuclear physics, electronics, materials, semiconductors, solid-state physics, theoretical physics, radiation physics, activation analysis, astronomy, polymer physics, and biophysics among them.

Uzbekistan women scientists are successfully working in almost all of these fields, including:

- Mukhae Rasulova – mathematical physics (Bogolyubov's equations)
- Manzura Usmanova – analytical activation analysis of high-pure materials
- Ella Ibragimova – solid-state radiation physics and high-temperature superconductors
- Feruza Umarova – physics of solid-state theory: calculation of electronic defect structure and the defect's influence on properties of Si semiconductor materials
- Galina Ni – nuclear reaction theory
- Dilbar Gulamova – solar materials physics
- Ada Liderman – semiconductor physics

At present in Uzbekistan there are more than 20 high schools with physics departments with high-quality curricula preparing researchers in various physical science disciplines. Laboratory research is conducted in different fields of physics. For example, under Sayera Rashidova's leadership at National University (and at the Polymer

CP628, *Women in Physics: The IUPAP International Conference on Women in Physics,* edited by B. K. Hartline and D. Li

Chemistry-Physics Institute) investigations are conducted on modification and properties of polymer materials. The team of Ainisa Tashmukhamedova at National University is investigating the structure and properties of crown ether and its ion complexes.

A group of women researchers under Marina Zamaraeva's leadership in the Biophysics Department of National University is working on membrane enzymes structure and its properties. In the Physics Department of Samarkand University, Maysara Salakhitdinova and other women researchers are working in the field of atoms and molecular spectroscopy.

CONTEMPORARY CONDITIONS AND TENDENCIES

In recent years, because of experimental equipment becoming out of date, some scientists have changed their field of investigation to theoretical research and modeling. Because of insufficient financial support of science by the government, and a low standard of living, many researchers are forced to find additional work to support their families.

The rate of young researchers entering physics has essentially decreased after the demise of the former Soviet Union and the introduction of a market system. Now some of our junior persons are studying in Western countries, but mainly in the humanitarian and marketing areas. Most young people prefer to go into the money professions.

In the last few years some research teams have received international grants, which has served to promote scientific research in our country, attracting some young people, as well as older, more experienced and capable women researchers.

Participation of Women
in the Development of Physics in Yugoslavia

Maja Burić[1], Agika Kapor[2], Dragana Popović[3], Mirjana Popović-Božić[4],
and Mirjana Vuceljić[5]

[1]University of Belgrade; [2]University of Novi Sad; [3]Faculty of Veterinary Medicine, University of Belgrade;
[4]Institute of Physics; [5]University of Montenegro

The textbook *Fysika* by Atanasije Stojković (professor of physics in Kiev), published in Budim, in 1801/1803, marks the beginning of the rise of physics among nations living in today's Yugoslavia. The book was inspired by Dositej Obradović, one of the most educated persons in the region, who made a great effort to start an education system for Serbs and to improve the social status of women. In one of his papers he wrote that "it would be great if certain Serbs would undertake writing physics for slaveno-serb people."

In 1808, Dositej founded the first Serbian school, "Velika škola." After 1863, Velika škola included the Faculty of Law, the Philosophical Faculty, and the Technical Faculty, which offered physics courses. Still, in order to get university diplomas, Serbs had to go to universities in Europe. In the first decade of the 20th century at European universities, among 120 women students from the Balkans, 19 were Serbs. Among them was Mileva Marić, the first wife of Albert Einstein, whose influence/contribution to his work is still a subject of controversy.

The University of Belgrade was founded in 1905. Physics was studied at the Faculty of Philosophy within the Chair of Natural Sciences. Physics Chairs were formed at the Technical Faculty and at the Faculty of Medicine, offering courses such as physics, technical physics, and medical physics. Still, there is no record of the participation of women in university physics education before the Second World War.

The new social system established after the war proclaimed the importance of the participation of women both in professions and in social activities. Yugoslav women voted for the first time in 1945. The new Constitution (1946) gave women the right to vote and proclaimed equal rights and opportunities to women and men.

In 1946, Dragica Nikolić and Branka Radivojević became assistants of physics at the Faculty of Philosophy, and Dragica Kirić at the Faculty of Medicine. Dragica Nikolić was a delegate at the 1st Congress of Mathematicians and Physicists of FNRYugoslavia, held in 1949. Slowly, more women enrolled in physics at the faculties of Natural Sciences founded in Belgrade (1947), Novi Sad (1961), Kragujevac (1972), Niš (1974), and Podgorica (1985).

Many women interested in physics and its technical applications have studied at the Department of Engineering Physics (now the Department of Physical Electronics), founded in 1955 at the Faculty of Electrical Engineering, University of Belgrade. Physicists, both men and women, who received their bachelor's and Ph.D.'s used to get jobs in public schools, faculties, and research institutes, but many of them, both men and women, went abroad to the U.S. and Western Europe. The number of physicists who flew abroad increased significantly during the last decade.

The Institute of Nuclear Sciences Vinča, founded in 1948, and the Institute of Physics in Belgrade, founded in 1961, have attracted women physicists from the very beginning. Mira Jurić and Branislava Perović played significant roles in the development of nuclear methods and instrumentation in the Institute Vinča.

Mira Juric, starting as assistant in 1948, became the leader of the group studying nuclear reactions by analyzing particle tracks on photo-emulsions. She became professor at the Faculty of Natural Sciences and founded a new group for particle detectors in the Institute of Physics in Belgrade. She paved the way for groups from the Institute Vinča and the Institute of Physics to participate in international collaborations in elementary particle physics.

The development of physics of ionized gases was initiated in 1965 and led for many years by Branislava Perović, who taught at the Department of Engineering Physics. The physics of ionized gases is an active field of research today in Yugoslavia, characterized by intensive international collaboration. From 1976 to 1979, Branislava Perović was the Head of the Vinča Institute.

CP628, *Women in Physics: The IUPAP International Conference on Women in Physics,* edited by B. K. Hartline and D. Li
© 2002 American Institute of Physics 0-7354-0074-1/02/$19.00

TABLE 1. Average number of Physics and Technical Physics Bachelor's Degrees at Six Yugoslav Institutions During the Past 15 Years.

Faculty/ University*	Female		Male		
	Total	Avg/yr	Total	Avg/yr	% women
FNSM UB/ FPh UB	274	18.27	331	22.07	45.3 %
FNSM UNS	94	6.27	75	5.00	55.6 %
FNSM UM	53	3.53	28	1.87	65.4 %
FNSM UK	80	5.33	31	1.94	72.0 %
FNSM UN	60	4.00	60	4.00	50.0 %
FEE UB	197	13.13	444	29.6	30.7 %
Totals	758	50.53	969	64.6	43.9 %

* FNSM UB / FPh UB – Faculty of Natural Sciences and Mathematics/ Faculty of Physics, University of Belgrade; FNSM UNS – Faculty of Natural Sciences and Mathematics, University of Novi Sad; FNSM UK – Faculty of Natural Sciences and Mathematics, University of Kragujevac; FNSM UN – Faculty of Natural Sciences and Mathematics, University of Niš; FEE UB – Faculty of Electrical Engineering, Department of Physical Electronics, University of Belgrade.

The Laboratory for Atomic Physics, founded at the Institute of Physics in the 1960s by Prof. M. Kurepa, has had until now about 20 women physicists. The Laboratory gained an international reputation in the field of electron collisions with atoms or molecules. At present about a dozen former members of this Laboratory, mostly women, are scientists and professors at leading universities in Australia, Belgium, the UK, France, Germany, Slovenia, Sweden. and the United States.

Leposava Vusković, among them, has very successfully combined a scientific and pedagogical career working at the University of New York, at Old Dominion University in Norfolk, and on NASA and NSF projects.

Gravitation, field theory, physics of elementary particles, quantum mechanics, mathematical physics, group theory, atomic physics, condensed state, statistical physics, quantum optics, and nonlinear optics are today's fields of active theoretical research in Yugoslavia. LjiLjana Dobrosavljević-Grujić has contributed much to the development of theoretical solid-state physics. She received her Ph.D. degree from the Ecole Normale Supérieure with P.G. de Gennes. Her research in superconductivity has attracted a great number of M.Sc. and Ph.D. students. Many of them continued their careers in the United States.

From the historical survey and statistical data (Tables 1–3) we might conclude that women in Yugoslavia began participating in physics as soon as it was socially permitted. The participation of women today is quite significant.

Women physicists in Yugoslavia do not feel any gender discrimination today. This is due to the Yugoslav constitution and laws, the general attitude toward women, and family tradition. We point out that a leave of absence for pregnancy and childbirth was six months before 1978, but it has been one month plus one year since than. This has allowed to women physicists in Yugoslavia to build a family as well as a professional career.

TABLE 2. Physics M.Sci. and Ph.D. Degrees Awarded at Four Yugoslav Institutions Since Their Inception.

Faculty University	Period	M.Sci.				Period	Ph.D.			
		Female		Male			Female		Male	
		Total	Avg/yr	Total	Avg/yr		Total	Avg/yr	Total	Avg/yr
FNSM UB FPh UB	1947-2001	124	2.3	347	6.4	1947-2001	65	1.2	226	4.2
FNSM UNS	1985-2000	2	0.15	2	0.1	1972-2000	10	0.7	26	1.85
FNSM UK	1984-2000	4	0.2	7	0.4	1981-2001	1	0.1	8	0.5
FNSM UN	1971-2000	1		4	0.1	1971-2001	0	0	6	0.2

TABLE 3. Women/men physicists in Yugoslav research institutes in 2001.

University	Research Trainee		Assistant		Research Assistant		Research Associate		Research Professor		Total	
	F	M	F	M	F	M	F	M	F	M	F	M
Institute of Physics, Belgrade	3	6	16	19	8	22	3	7	3	17	33	71
Institute for Nuclear Sciences, Vinča	8	19	8	15	4	18	3	12	3	20	26	84

Women in Physics in Zimbabwe

Sekai Shambira

Medical Physics Department, Johannesburg Hospital, South Africa

There are very few women pursuing physics careers in Zimbabwe for a number of reasons. The major drawbacks are outlined below, not in any particular order. Also outlined at the end are suggestions on how this issue can be solved.

There has traditionally been a preconceived idea that technical science subjects, especially physics, are for men, and that women are not fit to study them. While men are automatically acknowledged and accepted in the physics field, women's abilities are doubted until proven and they therefore start with a disadvantage. They then have to work twice as hard as men to get the same recognition. Women are perceived as less talented, less authoritative, and less capable. In view of all this, it is important to increase the participation of women in physics in my country so the issue of imbalance between men and women participants can be addressed and this traditional myth dispelled.

Most girls are raised with low professional expectations, and instead are focused on getting married and being a housewife and looking after children. They are discouraged from considering science or engineering careers. Once you join the field of physics, you are on your own. The men do not want to associate with you because you are "invading their territory" and they feel threatened. This is more prevalent in coed schools than in all girls' schools. The women think you want to be a man so they do not really want to associate with you.

The most serious barriers to women achieving successful physics careers in Zimbabwe are the lack of career guidance and mentoring, and the discouragement and scaring off of prospective physics students by society and financial constraints.

In most areas of Zimbabwe (especially rural) the average family is made up of appreciably more than two children and, with the harsh economic conditions parents cannot afford to send all the children to school. Our culture is such that the man is the head of the house and should be able to support the family while the woman does the household chores. Financial constraints result in the boy child getting the education while the girl child remains at home or is even married off early to pay the brother's school fees. In the rare event that they both go to school, the girl child is expected (culturally) to do the household chores before and after school. After such a long, hard day it is not surprising that she cannot give as much effort and concentration as is required to do well in physics (or any demanding subject) as a boy child. She therefore ends up concentrating in the arts (easy to read while doing some household chore, such as preparing a meal) and the field of physics is deprived of yet another woman.

To alleviate these barriers in my country, physicists should help encourage girls to study physics. They should organize seminars and talks at schools, with speakers preferably being women physicists who have made it to the top. We need role models who can show intelligent girls that they can pursue a physics career. Most girls in Zimbabwe, living in a less industrialized third-world country, do not know of the opportunities that exist for physicists. It is important to discuss these with them so that they can see there is a future in a physics career.

In Zimbabwe, much has to be done to attract women, to tell them, for example, that they can still start their own industries and/or work in the existing industries. In short, career guidance is of the utmost importance. Exchange programs for high school students should also be organized during holidays. Assistance (financial or otherwise) from other countries can also help women attain higher qualifications in physics. For example, I am undergoing postgraduate training in medical physics at the Johannesburg Hospital in South Africa, and my colleague Jane Gore was sponsored by the National Science Foundation to pursue her Ph.D.

Organizations such as the Zimbabwe Association of University Women sponsor financially disadvantaged girls to pursue a general education. We need organizations such as the Third World Organization for Women in Science, (TWOWS), in physics. Unfortunately, most of these existing organizations offer only postgraduate assistance, whereas the greatest need is at the high school and undergraduate level of the women's education.

CP628, *Women in Physics: The IUPAP International Conference on Women in Physics,* edited by B. K. Hartline and D. Li
© 2002 American Institute of Physics 0-7354-0074-1/02/$19.00

The Zimbabwean government has introduced affirmative action where women applicants to universities and other institutions of higher learning are considered using lower entry qualifications than their male counterparts. This initiative, though it looks like it undermines women's abilities, is aimed at equalizing opportunity and seems to have increased the number of women enrolled in physics during the short time that it has been in existence.

There is also a need for a network for women physicists to interact and share experiences. We should learn to be assertive and demand recognition for our efforts without turning this into a gender-based issue.

From this conference, we hope to get clues on how to encourage girls to study physics, how to balance family with a demanding physics career without compromising one or the other, and how to overcome the above-mentioned barriers to women in the field of physics.

APPENDICES

Appendix A
Conference Program

March 7

9:00–10:30	**Plenary Session 1:** **Chair: Marcia Barbosa**, Universidade Federal do Rio Grande do Sul, Brazil; Chair, IUPAP Working Group on Women in Physics
9:00–9:30	**Welcome Remarks From:** **Burton Richter**, President of IUPAP **W. Erdelen**, (Assistant) Director-General for Natural Sciences (UNESCO) **Philippe Busquin**, Commissioner for Research of the European Union
9:30–10:00	**Roman Czujko***, Director of the AIP Statistical Research Center: "Resources, Opportunities and Encouragement: Findings from the International Study of Women in Physics"
10:00–10:30	**Teresa Rees***, Cardiff University of Social Sciences, UK: "Women and Science in Europe: A Review of National Policies"
10:30–11:00	**Morning Break**
11:00–12:30	**Plenary Session 2:** **Chair: Oliviu Gherman**, Ambassador to France, Romania
11:00–11:30	**Claudine Hermann***, Deputy Director, Laboratory of Physics of Condensed Matter, Ecole Polytechnique, France: "The European Union Report on Women and Science and a French Experience"
11:30–12:00	**Zhang Xuezhong**, Ambassador to UNESCO, China-Beijing, on behalf of Chen Zhili, Minister of Education, China-Beijing: "Women in Physics: The View From China"
12:00–12:30	**Karimat Mahmoud el-Sayed***, Vice-Dean, Cairo University, Egypt: "Women in Physics: The Situation in Egypt"
12:30–14:00	**Lunch**

*All plenary talks will be 25 minutes to allow 5 minutes for questions and discussion.

Conference Program (continued)

14:00–15:30	**Small Group Discussions 1** • Attracting Girls Into Physics • Launching a Successful Physics Career • Getting Women Into the Physics Leadership Structure Nationally and Internationally
15:30–16:00	**Afternoon Break**
16:00–17:30	**Small Group Discussions 1** (continued)
17:30–19:30	**Poster Session 1 and Cocktail Reception**

March 8

9:00–10:30	**Plenary Session 3:** **Chair: Hidetoshi Fukuyama,** University of Tokyo, Japan; Member of IUPAP Working Group on Women in Physics
9:00–9:30	**Elisa Baggio Saitovitch***, Vice-President of the Brazilian Physical Society, Brazil: "Personal Experience as a Latin American Physicist"
9:30–10:00	**Masako Bando***, Aichi University, Japan: "Status of women in Physics of Japan and Future Aspects: Findings From Questionnaire of JPS and JAPS"
10:00–10:30	**I.P. Ipatova***, A. F. Ioffe Institute, Russia: "Russian Women: Line of Life"
10:30–11:00	**Morning Break**
11:00–12:30	**Plenary Session 4:** **Chair: Elisa Molinari,** University of Modena, Italy; Member of IUPAP Working Group on Women in Physics • 11:00: Report from Topic 1: Samathi Rao, India; Member of IUPAP Working Group on Women in Physics • 11:30: Report from Topic 2: Beverly Hartline, Argonne National Laboratory, USA; Member of IUPAP Working Group on Women in Physics • 12:00: Report from Topic 3: Katharine Gebbie, NIST, USA; Member of the IUPAP Working Group on Women in Physics
12:30–14:00	**Lunch**

Conference Program (continued)

14:00–15:30	**Small Group Discussions 2** • Improving the Institutional Climate for Women in Physics • Learning From Regional Differences • Balancing Family and Career
15:30–16:00	**Afternoon Break**
16:00–17:30	**Small Group Discussions 2** (continued)
17:30–19:00	**Poster Session 2**
19:00–	**Departure for Dinner/Boat Trip on Seine**

March 9

9:00–10:30	**Plenary Session 5:** **Chair: Elisabeth Giacobino,** Director, Department of Physics and Mathematics Sciences, CNSR, France (invited)
9:00–9:30	**Catherine Cesarsky***, Director General of the European Southern Observatory: "Women in Science: Personal Impressions"
9:30–10:00	**Nancy Hopkins***, Amgen, Inc., Professor of Biology, MIT, USA: "Women Faculty in Science at MIT"
10:00–10:30	**Rohini Godbole***, Chair, Center for Theoretical Studies, Indian Institute of Science, India: "Being a Women Physicist: An Indian Perspective"
10:30–11:00	**Morning Break**
11:00–12:30	**Plenary Session 6:** **Chair: Marcelle Rey-Campagnolle,** France, Chair of the Organizing Committee • 11:00: Report from Topic 4: Ling–An Wu, Institute of Physics, China; Member of IUPAP Working Group on Women in Physics • 11:30: Report from Topic 5: Pia Thörngren-Engblom, Uppsala University, Sweden • 12:00: Report from Topic 6: Barbara Sandow, Free University of Berlin, Member of IUPAP Working Group on Women in Physics
12:30–14:00	**Lunch**
14:00–16:00	**Plenary Session 7:** **Chair: Judy Franz,** Associate Secretary General, IUPAP, USA
14:00–15:00	**Presentation of Resolutions and Voting on Resolutions**
15:00–15:30	**Voting on Conference Resolutions**
15:30–16:00	**Conference Follow–Up**
16:00–16:15	**Adjourn**

Appendix B
Conference Participants

An asterisk indicates a country's team leader for the conference.

Albania	*Antoneta Deda, *antonetad@yahoo.com*; *adeda@fshn.tirana.al* Vera Bekteshi, *verabek2000@yahoo.com* Odeta Çati, *okojaal@yahoo.com*
Argentina	*Silvina P Dawson, *silvina@dfuba.df.uba.ar* Karen Hallberg, *karen@cab.cnea.gov.ar*
Armenia	*Inna Aznauryan, *aznaur@jerewanl.yerphi.am* Alita Danagulyan, *danag@ysu.am* Nina Demekhina, *nina@1xyerphi.am*
Australia	Giuseppina Dall"Armi-Stoks, *giuseppina.dallarmi-stoks@dsto.defence.gov.au* Manjula Sharma, *m.Sharma@physics.usyd.eud.au*
Austria	*Claudia Ambrosch-Draxl E-mail, *Claudia.ambrosch@uni-graz.at* Verena Grill, *verena.grill@uibk.ac.at* Monika Ritsch-Marte, *monika.ritsch-marte@uibk.ac.at* Kerstin Weinmeier, *kerstin.weinmeier@uni-graz.at*
Belarus	*Larissa Svirina, *lsvirina@dragon.bas-net.by* Zalesskaya Galina Adamovna, *zalesskaya@imaph.bas-net.by* Iryna Miadzuedz (contact through team leader) Ershov-Pavlov Evgenii Anatol'evich, *ershov@imaph.bas-net.by*
Belgium	*Petra Rudolf, *petra.rudolf@fundp.ac.be* Nathalie Balcaen, *Nathalie.balcaen@rug.ac.be* Peggy Fredrickx, *pefred@ruca.ua.ac.be* Christine Iserentant, *Christine.iserentant@rug.ac.be* Griet Janssen, *grietj@uia.ua.ac.be* Karen Janssens, *karenj@uia.ua.ac.be* Muriel Vander Donckt, *mvd@iba.be*
Botswana	*S.F. Mpuchane, *mpuchans@mopipi.ub.bw* (was not present) Dorcus Tau, *Fax 267 320423* Bogodile Matlhape, *Fax 267 373338*

Conference Participants (continued)

Brazil	*Marilia Caldas, *mjcaldas@usp.br* Marcia Barbosa, *barbosa@if.ufrgs.br* Solange Cavalcanti, *solange@lux.ufal.br* Yvonne Primerano Mascarenhas, *Yvonne@if.sc.usp.br* Elisa Saitovitch, *elisa@cbpf.br*
Bulgaria	*Ana Proykova, *anap@phys.uni-sofia.bg* Simona Kouteva, *kouteva@tu-cottbus.de* Penka Lazarova, *science@bitex.com* Zhelyazka Raykova-Bozova, *janeraik@pu.acad.bg*
Cameroon	*Ndukong Tata Gerard, *alliedbda@bdanet.cm* (was not present) Samba Odette Ngano (contact through team leader)
Canada	*Marie D'Iorio, *marie.d'iorio@nrc.ca* Janis McKenna, *janis@physics.ubc.ca* Ann McMillan, *ann.mcmillan@ec.gc.ca* Eric Svensson, *eric.svensson@nrc.ca*
China-Mainland	*Ling-An Wu, *wula@aphy.iphy.ac.cn* Manlin Sui, *mlsui@imr.ac.cn* Shu-qin Tian, *xieyicheng@yeah.net* Ming Wang, *zqxi@16net* Yicheng Xie, *xieyicheng@yeah.net* Yanlai Yan, *ylyan@online.sh.ch* Hong Zhang
China-Taiwan	*Ming-Fong Tai, *phymft@ccunix.ccu.edu.tw* Yeong-Chuan Kao, *yckau@phy.ntu.edu.tw* Jauym Grace Lin, E-mail, *jglin@phys.ntu.edu.tw* Li-Ling Tsai, *ltsai@interchange.ubc.ca* Hue Min Wu, *hueminwu@faculty.pccu.edu.tw* Maw-Kuen Wu, *mkwu@phys.nthu.edu.tw*
Colombia	*Angela Camacho, *acamacho@uniandes.edu.co*
Croatia	*V. Lopac, *vlopac@marie.fkit.hr* Planinka Pecina, *planinka@phy.hr* A. Tonejc *andelka@phy.hr*
Cuba	*Lilliam Alvarez Diaz, *lilliam@cidet.icmf.inf.cu* Maria Margarita Cobas Aranda, *margarita@aen.energia.inf.cu*
Czech Republic	*Raji Heyrovska, *Raji.Heyrovska@jh-inst.cas.cz* Jarmila Kodymova, *kodym@fzu.cz* Olga Krupkova, *olga.krupkova@math.slu.cz* Jana Musilova, *janam@physics.muni.cz*

Conference Participants (continued)

Denmark	Nils O. Andersen, *noa@fys.ku.dk*
	Cathrine Fox Maule, *foxmaule@gfy.ku.dk*
	Cathrine Hasse, *ch@humanities.dk*
	Liv Hornekaer, *hornekaer@fysik.sdu.dk*
	Henriette Jensenius, *hj@mic.dtu.dk*
	Birgitta Nordstroem, *birgitta@astro.ku.dk*
Egypt	Karimat Mahmoud El-Sayed, *karima@mailer.scu.eun.eg*
	Seham Ahmed Abd-El-Hady, *saabdelhady@hotmail.com*
Estonia	*Helle Kaasik, *helle@eeter.fi.tartu.ee*
	Vladimir Hizhnyakov, *hizh@fi.tartu.ee*
	Imbi Tehver, *tehver@fi.tartu.ee*
	Anu Ylejoe (Ulejoe), *anu.ulejoe@raad.tartu.ee*
Finland	*Helena Aksela, *Helena.aksela@oulu.fi*
	Kukka Banzuzi, *kukka.banzuzi@cern.ch*
	Jonna Koponen, *jmkopone@rock.helsinki.fi*
	Ulla Lahteenmaki, *ulla.lahteenmaki@mikes.fi*
	Anna Penttila, *anna.penttila@oulu.fi*
France	*Claudine Hermann, *claudine.hermann@polytechnique.fr*
	Christian Bastain, *c.bastian@univ-mulhouse.fr*
	Violette Brisson, *brisson@lal.in2p3.fr*
	Catherine Cesarsky, *adellerb@eso.org*
	Alain Costes, *www.recherche.gouv.fr/technologie/mission/default.htm*
	Vesna Cuplov, *cuplov@cpt.univ-mrs.fr*
	Nicole Dewandre, *Fax 32 2 299 49 25*
	Martial Ducloy, *ducloy@galilee.univ-paris13.fr*
	W. Erdelen, *werdelen@unesco.org*
	Elisabeth Giacobino, *Elisabeth.giacobino@cnrs-dir.fr*
	H. Godfrin, *godfrin@labs.polycnrs.gre.fr*
	Etienne Guyon, *Etienne.guyon@ens.fr*
	David Lee, *d.lee@uha.fr*
	Martine Lumbreras, *lumbre@ese-metz.fr*
	Gen. Koichiro Matsuura, *kmatsuura@unesco.org*
	Alarcon Minella, *m.alarcon@unesco.org*
	Marcelle Rey-Campagnolle, *m.rey-campagnolle@wanadoo.fr*
	Monique Schwob, *monque.schwob@wanadoo.fr*
	Annick Suzor-Weiner, *annick.suzor-weiner@ppm.u-psud.fr*
	Zhang Xuezhong, *dl.chin2@unesco.org*

Conference Participants (continued)

Germany	*Barbara Sandow, *Sandow@physik.fu-berlin.de* Silke Bargstaedt-Franke, *Silke.bargstaedt-franke@infineon.com* Monika Bessenrodt-weberpals, *mob@ipp.mpg.de* Alexander Braddshaw, *alex.bradshaw@ipp.mpg.de* Carmen Espinola, *cespinol@mail.uni-mainz.de* Stephanie Hastmann, *s_hastmann@web.de* Corinna Kausch, *c.kausch@gsi.de* Susanne Metzger, *Susanne.Metzger@uni-mainz.de* Josifina Sarrou, *sarrou@spin.chem.tu-berlin.de*
Ghana	*Aba Andam, *abaandam@yahoo.com* Paulina Amponsah, *pekua2@hotmail.com* Elsie Kaufman, *abaandam@yahoo.com*
Greece	Christine Kourkoumelis, *hkourkou@cc.uoa.gr* Eleni Pavlidou, *pavlidou@physics.auth.gr* Charclia Petridou, *petridou@mail.cern.ch* Christina Solomonidon, *xsolom@uth.gr* Eleni Stavidou, *estavrid@uth.gr* Ekaterini Tsoulou, *ktsoulou@mail.demokritos.gr*
Hungary	*Judit Nemeth, *judit@poe.elte.hu* Zsuzsanna Gyory, *gyory@complex.elte.hu* Agnes Vibok, *vibok@cseles.atomki.hu*
India	Rohini Godbole, *rohini@cts.iisc.ernet.in* Neelima Gupte, *gupte@chaos.iitm.ernet.in* Jyoti Gyan Chaudhuri, *gjyoti@apsara.barc.ernet.in* Sunita Nair, *sunita@rri.res.in* Sumathi Rao, *sumathi@mri.ernet.in*
Indonesia	*Wiwik S. Subowo, *wiwik@p3ft.lipi.go.id*; *p3ftlipi@bdg.centrin.net.id* Frida Ulfah Ermawati, *frida1@telkom.net* Anung Kusnowo, *anung@mss.lipi.go.id*
Iran	*Azam Iraji-zad, *iraji@sina.sharif.ac.ir* Mahboubeh Houshiar, *m-houshiar@cc.sbu.ac.ir* Azam Pourghazi, *azpour@yahoo.com*
Ireland	Aine Allen, *Aine.Allen@it-tallaght.ie* Sue McGrath, *suemcgrath@w5online.co.uk* John O'Brien, *john.obrien@ul.ie*
Israel	*Itzhak Tserruya, *itzhak.tserruya@weizmann.ac.il* Halina Abramowicz, *holina@post.tav.ac.il* Sara Beck, *sara@wise.tau.ac.il* Yael Shadmi, *yshadmi@wicc.weiszmann.ac.il*

Conference Participants (continued)

Italy	*Elisa Molinari, *molinari@unimo.it* Maria Grazia Betti, *betti@roma1.infn.it* Annalisa Bonfiglio, *annalisa@unica.it* Hilda Cerdeira, *cerdeira@ictp.trieste.it* Maria Fittipaldi, *maria@molphys.leidenuniv.nl* Daniela Grasso, *daniela@delphi.polito.it* Anna Grazia Mignani, *mignani@iroe.fi.cnr.it* Maria Antonietta Loi, *antonietta.loi@jku.at* Maria Luigia Paciello, *mariella.paciello@roma1.infn.it*
Japan	*Kazuo Kitahara, *kazuo@icu.ac.jp* Masako Bando, *bando@aichi-u.ac.jp* Hidetoshi Fukuyama, *fukuyama@issp.u-tokyo.ac.jp* Akiko Gomyo, *gomyo@frl.cl.nec.co.jp* Machiko Hatsuda, *mhatsuda@post.kek.jp* Yuhri Ishimaru, *ishimaru@iap.fr* Kazue Ishikawa, *kazue-I@hoffman.cc.sophia.ac.jp* Kayoto Ito, *ito@jsap.or.jp* Tomoko Kagayama, *kagayama@kumamoto-u.ac.jp* Kashiko Kodate, *kodate@fc.jwu.ac.jp* Yukari Matsuo, *ymatsuo@riken.go.jp* Izumi Nomura, *nomura@msnifs.ac.jp* Madoka Takai, *takai@micro.mm.t.u-tokyo.ac.jp* Emi Tamechika, *emi@aecl.ntt.co.jp*
Korea	*Kwang-Hwa Chung, *khchung@kriss.re.kr* Young Soon Kim, *yskim@mju.ac.kr* Jeong Won Wu, *jwwu@ewha.ac.kr*
Latvia	*Erna Karule-Gailite, *karule@latnet.lv* , *gailitis@sal.lv* Inta Muzikante, *intam@edi.lv* Gita Revalde, *gitar@latnet.lv*
Lithuania	*Alicija Kupliauskiene, *akupl@itpa.lt*, Rasa Kivilsiene, *rasa@itpa.lt* Dalia Shatkovskiene, *dalia.satkovskiene@ff.vu.lt* Saule Vingeliene, *saulving@centras.lt*
Malaysia	*Khalijah Salleh, *khalijah_s@hotmail.com* Samirah Abdul Rahman, *sam_are@pop.jaring.my* Azni Zain-Ahmed, *azain_ahmed@hotmail.com*
Mexico	*Lilia Meza Montes, *lilia@sirio.ifuap.buap.mx* Ana Maria Cetto, *ana@fisica.unam.mx* Xochitl Lopez Lozano, *xlopez@sirio.ifuap.buap.mx*
Netherlands	*Helene M. van Pinxteren *helene@rijnh.nl* Annalisa Fasolino *fasolino@sci.kun.nl* Eddy W.A. Lingeman *ed@nikhef.nl* Christa Hooijer, *christa.hooijer@fom.nl*

Conference Participants (continued)

New Zealand	*Jenni Adams, *j.adams@phys.canterbury.ac.nz* Pauline Harris, *p.harris@phys.canterbury.ac.nz*
Nigeria	*Ibiyinka A. Fuwape, *yfuwape@yahoo.com* Oyebola Popoola, *oluwase@skannet.com.ng*
Norway	*Ashild Fredriksen, *ashild.fredriksen@phys.uit.no* Anne Borg, *anne.borg@phys.ntnu.no*
Pakistan	*Rashid Khalid, *astrophy@comsats.net.pk* Aziz Fatima Hasnain, *nazr@khi.compol.com* Asghar Maasood, *tpl.qau@usa.net*
Poland	*Izabela Sosnowska, *izabela@fuw.edu.pl* Boguslawa Adamowicz, *adamowic@zeus.polsl.gliwice.pl* Maria Giller, *mgiller@kfd2.fic.uni.lodz.pl* Teresa Rzaca-Urban, *rzaca@fuw.edu.pl*
Portugal	*Ana Maria Eiro, *ana.eiro@at.fc.ul.pt*
Romania	*Anca Visinescu, *avisin@theor1.theory.nipne.ro* Sanda Adam, *adams@theor1.theory.nipne.ro* Agneta Balint, *sbalint@mail.dnttm.ro*; *balint@quasar.physics.uvt.ro* Florenta Costache, *costache@tu-cottbus.de* Violeta Georgescu, *vgeor@uaic.ro* Oliviu Gherman, *senatro@dias.vsat.ro* Maria Giubelan, *m_giubelan@chim.upb.ro*
Russia	*Galina Merkulove, *merk.gal@pop.ioffe.rssi.ru*; *merk@ustu.neva.ru* Nelly Didenko, *didenko@spbrc.nw.ru* Iya Ipatova, *iip.ton@pop.ioffe.rssi.ru* Svetlana Markova, *snm@sai.msu.ru* Eduard Tropp, *tropp@spbrc.nw.ru*
Senegal	*Faye Ndeye Arame Boye, *akonte@ucad.sn* Thiandaime Coumba (contact through team leader)
Slovak Republic	*Eva Majkova, *majkova@savba.sk* Anna Dubnickova, *dubnickova@fmph.uniba.sk* Adela Kravcakova, *kravcak@hep.science.upjs.sk* Marcela Morvova, *morvova@fmph.uniba.sk*
Slovenia	Andreja Gomboc, *andreja@fiz.uni-lj.si* Nada Razpet, *nada.razpet@guest.arnes.si* Maja Remskar, *maja.remskar@ijs.si*
South Africa	*Jaynie Padayachee, *padayachee@nac.ac.za* Mmantsae Diale, *mdiale@postino.up.ac.za* E.C. Viljoen, *elmarie@tlabs.ac.za*

Spain	*Mã Josefa Yzuel, *maria.yzuel@uab.es*
	Carmen Carreras, *ccarreras@ccia.uned.es*
	Margarita Chevalier, *chevalier@eucmax.sim.ucm.es*
	Teresa Lopez, *mteresa.lopez@iberdrola.es*
	M. Pilar Lopez Sancho, *Pilar@icmm.csic.es*
	Florencia Penna, *flor.penna@uam.es*
Sudan	*Osman Mai Eltag Mohamed, *maieltag@hotmail.com*
	Abdelrazig Mohamed Abdelbagi, *razig2000@hotmail.com*
	Azza Abdelrahim Mohamed Ahmed, *azza_rahim@yahoo.com*
Sweden	*Katarina Wilhelmsen Rolander, *wilhelmsen@physto.se*
	Lotten Glans, *lotten.glans@mh.se*
	Karoline Wiesner, *karoline.wiesner@fysik.uu.se*
	Ulla Tengblad, *fpdd@tsl.uu.se*
	Pia Thörngren, *pia.thorngren@tsl.uu.se*
Switzerland	*Iris Zschokke, *iris.zschokke@unibas.ch*
	Christina Matthey, *Christina.matthey@cern.ch*
	Felicitas Pauss, *felicitas.pauss@cern.ch*
	Michela Rossi, *michela.rossi@ito.umnw.ethz.ch*
	Silvia Schuh, *silvia.schuh@cern.ch*
Tunisia	*Zohra Ben Lakhdar, *zohra.lakhdar@fst.rnu.tn*
	Menana Kilani Gabsi, *menana_gabsi@hotmail.com*
	Souad Chekuir Lahmar, *souadl@altavista.com*
Turkey	*Engin Arik, *arik@boun,edu.tr*; *engin@ocean.phys.boun.edu.tr*
	Ayla Celikel, *ayurun@tr.net*
	Demet Kaya, *demet@itu.edu.tr*
	Saziye Ugur, *saziye@itu.edu.tr*
Ukraine	*Oksana Patsahan, *oksana@icmp.lviv.ua*
	Alla Moina, *alla@icmp.lviv.ua*
	Olena Vertsanova, *tempus@webber.net.ua*
United Kingdom	*Gillian Gehring, *g.gehring@sheffield.ac.uk*
	Yasmin Andrew, *yand@jet.uk*
	Joanne Baker, *jcb@astro.ox.ac.uk*
	Alex Byrne, *alex.byrne@iop.org*
	Sandra Chapman, *sandrac@astro.warwick.ac.uk*
	Dimitra Darambara, *dimitrad@medphys.ucl.ac.uk*
	Joanna Hamilton, *j.Hamilton@physics.gla.ac.uk*
	Helen Heath, *Helen.heath@bristol.ac.uk*
	Roisin Keenan, *keenan@qub.ac.uk*
	Ann Marks, *wipg@amarks.co.uk*

Conference Participants (continued)

United Kingdom (continued)	Yvonne Masakowski, *ymasakowski@onrifo.navy.mil* Peter Melville, *peter.Melville@iop.org* George Morrison, *g.c.Morrison@bham.ac.uk* Teresa Rees, *reestl@cardiff.ac.uk*
United States	*Meg Urry, *meg.urry@yale.edu* Jackie Beamon-Kiene, *beamon@aps.org* Kim Budil, *ksbudil@llnl.gov* Roman Czujko, *rczujko@aip.org* Judy Franz, *franz@aps.org* Katharine Gebbie, *kgebbie@nist.gov* Howard Gerogi, *georgi@physics.harvard.edu* Beverly Hartline, *bhartline@anl.gov* Nancy Hopkins, *nhopkins@mit.edu* JoAnn Joselyn, *jjoselyn@earthlink.net; sg@iugg.org* Kristine Lang, *kmlang@physics.berkeley.edu* Donqi Li, *dongqi@anl.gov* Laurie McNeil, *mcneil@physics.unc.edu* Frederick M. Pestorius, *mpestorius@onrifo.navy.mil* Burton Richter, *brichter@slac.stanford.edu* Erika Ridgway, *ridgway@aps.org* Peter Saeta, *saeta@hmc.edu* Jennifer Sokolowski, *jsokolos@cfa.harvard.edu* Sharon Stephenson, *sstephen@gettysburg.edu* Katie Stowe, *stowe@aip.org* Shelia Tobias, *Sheila@sheilatobis.com* Aparna Venkatesan, *aparna.venkatesan@colorado.edu*; *aparna@casa.colorado.edu* Judy Wyne, *wyne@onr.navy.mil* Yevgeniya Zastavker, *yzastavk@wellesley.edu*
Uzbekistan	*Ulmas Gafurov, *ulmas@suninp.tashkent.su; ulmasnew@yahoo.com* Firuza Umarova (contact through team leader)
Yugoslavia	Maja Buric, *majab@rudjer.ff.bg.ac.yu* Dragana Popovic, *stankopo@eunet.yu* Mira Vuceljic, *mirav@rc.pmf.cg.ac.yu*
Zimbabwe	*Shambire Sekai, *sekais@yahoo.com* Jane Gore, *jgore@science.uz.ac.zw*

Appendix C
List of Sponsors

IUPAP and the Conference Organizers thank the following sponsors, whose support and interest in women in physics made the conference and the associated activities possible.

International Union of Pure and Applied Physics

European Commission (5th Framework Programme for Research and Technological Development)

National Science Foundation (USA)

Department of Energy (USA)

International Council for Science (ICSU)

National Academy of Sciences (USA)

National Aeronautics and Space Administration (USA)

UNESCO Regional Bureau for Science in Europe

National Institute of Standards and Technology (USA)

American Physical Society

Institute of Physics (UK)

Centre National de la Recherche Scientifique (France)

The Physical Society of Japan

European Physical Society

French Ministry of Research

American Institute of Physics

Office of Naval Research International Field Office (USA)

Centre National de la Recherche Scientifique (France)

Japan Society of Applied Physics

European Science Foundation

European Physical Society

Engineering and Physical Sciences Research Council (UK)

Lawrence Livermore National Laboratory (USA)

Parliamentary Office of Science and Technology (UK)

International Centre for Theoretical Physics

KLA-Tencore (USA)

Particle Physics and Astronomy Research Council (UK)

Centro Latinoamericano de Fisica

AUTHOR INDEX

A

Abramowicz, H., 179
Adam, S., 211
Adams, J., 9, 199
Ahmed, A. Z., 193
Aksela, H., 159
Allen, A., 177
Alvarez Díaz, L., 151
Ambrosch-Draxl, C., 127
Andam, A., 9, 165
Andersen, A. C., 155
Arik, E., 13, 231
Aznauryan, I., 123

B

Baggio Saitovitch, E., 83
Baker, J., 21
Balcaen, N., 131
Balint, A., 211
Bando, M., 89
Barbosa, M. C., 135
Bargstaedt-Franke, S., 163
Beck, S., 179
Bessenrodt-Weberpals, M., 29, 163
Betti, M. G., 181
Bonfiglio, A., 181
Borg, A., 21
Budil, K. S., 17, 237
Burić, M., 241
Busquin, P., 45

C

Caldas, M., 13, 135
Camacho, A. S., 147
Carreras, C., 221
Celikel, A., 231
Cesarsky, C., 99
Cetto, A. M., 195
Chen, Z., 113
Chevalier, M., 221
Chung, K. H., 187
Cobas, M., 151
Costa, M. M. R., 209
Crespo, E., 221
Czujko, R., 49

D

Dall'Armi-Stoks, G., 125
Darambara, D., 13
Dawson, S. P., 121
Deda, A., 119
de la Viesca, R., 221
Didenko, N. I., 213
D'Iorio, M., 141
Dubničková, Z., 215

E

Eiró, A. M., 209
El-Sayed, K., 81
Erdelen, W., 43

F

Fasolino, A., 13, 197
Fredrickx, P., 21, 131
Fredriksen, Å., 9, 203
Fuwape, I. A., 201

G

Gabsi, M. K., 229
Gafurov, U., 239
García, M., 221
Gebbie, K. B., 17
Gehring, G., 13, 235
Georgescu, V., 211
Georgi, H., 237
Gerard, N. T., 139
Glans, L., 225
Godbole, R. M., 107
Gomboc, A., 217
Gupte, N., 9, 171
Guyon, E., 161
Gyanchandani, J., 9, 171

H

Haines, E., 199
Hallberg, K., 121
Hartline, B. K., 13
Hasnain, F., 205
Hermann, C., 75, 161
Heyrovska, R., 153
Hooijer, C., 9, 197

261

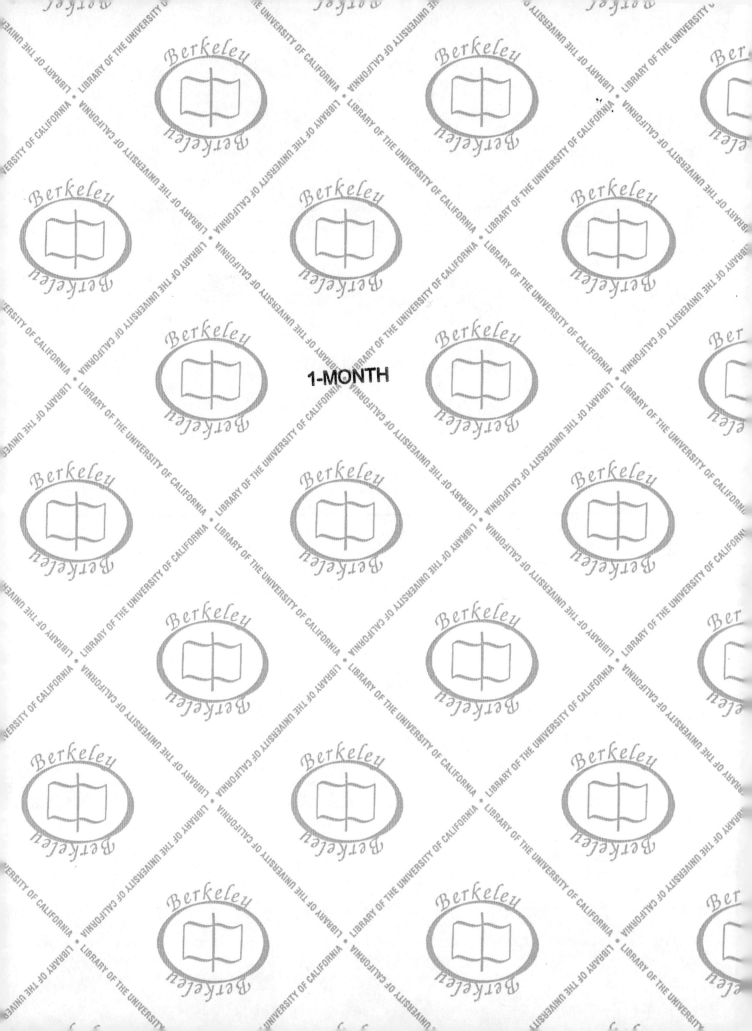